SOLID ACIDS AND BASES

Solid Acids and Bases

their catalytic properties

by

Kozo Tanabe

Professor of Chemistry, Hokkaido University,
Sapporo, Japan

1970

Kodansha, **Tokyo**

Academic Press, **New York • London**

Kodansha Scientific Books

Co-published by

Kodansha Ltd., 12–21 Otowa 2-chome, Bunkyo-ku, Tokyo 112, Japan

and

Academic Press Inc., 111 Fifth Avenue, New York, New York 10003, U.S.A.

United Kingdom Edition published by

Academic Press Inc. (London) Ltd., Berkeley Square House, London WIX 6BA, England

SALES TERRITORIES

Kodansha : Japan, Far East, S.E. Asia (east of and including W. Pakistan)

Academic Press: N. and S. America, Europe, Australasia, Africa, all other countries of Asia

LIBRARY OF CONGRESS CATALOG CARD NUMBER: 75-142084

INTERNATIONAL STANDARD BOOK NUMBER: 0-12-683250-1

KODANSHA EDP NUMBER: 3043-247036-2253 (0)

PRINTED IN JAPAN

Preface

Solid acids have found uses as catalysts for many important reactions including the cracking of hydrocarbons, the isomerization, polymerization and hydration of olefins, the alkylation of aromatics and the dehydration of alcohols, etc. Extensive investigations of solid acid catalysis over the last twenty years have resulted in major contributions to both fundamental research and industrial development, particularly in the field of petroleum chemistry. More recently, a number of new types of solid acid have been discovered and applied effectively as catalysts to a wide variety of chemical reactions.

Systematic study of the correlations between catalytic activity and selectivity and the acidic properties of the catalyst surface (the amount, strength and type—Brønsted or Lewis—of the acid sites) has, to a greater or lesser extent, enabled identification of the optimum catalyst in terms of these acidic properties. Why is it that seemingly neutral solid materials such as metal oxides, sulfates, etc. show remarkable acidic properties as well as catalytic activity? Elucidation is being achieved through investigations of the structure of the acid centres, and by comparison with the kinetics of homogeneous acid catalysis.

The comparatively little work that has as yet been performed on solid base catalysis gives promise that solid bases, too, will find further applications as effective catalysts for many important reactions. Both their catalytic activity and basic properties are currently being investigated by similar means to those for solid acid catalysts.

This monograph reviews recent developments in studies of the acidic and basic properties of solids, together with their catalytic activity and selectivity, and details the efficacy and special characteristics of solid acid and base catalysts.

It is hoped that this volume will be useful to both professional chemists and graduate students in the fields of organic, inorganic and physical chemistry, petroleum chemistry and catalysis, and also to all those with an interest in acidic and basic properties on solid surfaces.

I gratefully acknowledge my indebtedness to Dr. I. Matsuzaki, Dr. H. Hattori, Mr. R. Ohnishi and Mr. T. Yamaguchi—staff members of our laboratory—for their practical assistance and helpful discussions, and also to many other Japanese chemists, similarly engaged in research into solid acid and base catalysis, for their valuable suggestions and comments. The manuscript was typed by Miss H. Ikeda, whose assistance I acknowledge with thanks. I am grateful also to Miss Y. Yamada and Mr. R. D. Williams of Kodansha Ltd., who carefully checked and prepared the manuscript for publication; Mr. Williams duly emended the English text.

KOZO TANABE

Sapporo
June 1970

Department of Chemistry
Hokkaido University

Contents

Preface v
Acknowledgments viii

1. Solid Acids and Bases 1

2. Determination of Acidic Properties on Solid Surfaces 5

2.1 Acid strength 5
2.2 Amount of acid 13
2.3 Brønsted and Lewis acid sites 23
2.4 Relationship between acid strength and acid amount 28

3. Determination of Basic Properties on Solid Surfaces 35

3.1 Basic strength 35
3.2 Amount of base 38

4. Acid and Base Centres: Their Structure and Acid-Base Properties 45

4.1 Metal oxides and sulfides 45
4.2 Mixed metal oxides 58
4.3 Natural clays (zeolites, etc.) 73
4.4 Metal sulfates and phosphates 80
4.5 Others 90

5. Correlation between Acid-Base Properties and Catalytic Activity and Selectivity 103

5.1 Solid acid catalysis 103
5.2 Solid base catalysis 136
5.3 Solid acid-base bifunctional catalysis 145

6. Conclusion and Future Problems 159

Author index 163

Subject index 171

Catalyst index 174

Acknowledgments

Permission to reproduce the following copyright material is gratefully acknowledged.

Academic Press Inc., New York, "Advances in Catalysis"

Source	Used as
vol. 14, p. 143, Fig. 13 (H.P. Leftin and M.C. Hobson, Jr.)	Fig. 2–1
vol. 16, p. 60, A scheme of dehydration of menthol and neomenthol (H. Pines and J. Manassen)	Diagram (p. 149)
vol. 16, p. 64, A scheme of dehydration of *cis*, *cis*-1-decalol (H. Pines and J. Manassen)	Fig. 5–33

Editions Technip, Paris, "Actes du Deuxième Congrés International de Catalyse, Paris, 1961"

Source	Used as
Tome 1, No. 29, p. 714, Fig. 2 (J.B. Fisher and F. Sebba)	Fig. 5–3
Tome 2, No. 114, p. 2066, Fig. 4 (J.A. Rabo, P.E. Pickert, D.N. Stamires and J.E. Boyle)	Diagram (p. 74)

North-Holland Publishing Company, Amsterdam, "Proceedings of the Third International Congress on Catalysis"

Source	Used as
vol. 1, p. 401, Fig. 4 (M. Sato, T. Aonuma and T. Shiba)	Fig. 5–21
vol. 1, p. 417, Table 2 (M. Misono, Y. Saito, Y. Yoneda)	Table 5–4
vol. 1, p. 428, Fig. 5 (V. A. Dzisko)	Fig. 5–6
vol. 1, p. 429, Table 2 (V. A. Dzisko)	Table 5–5
vol. 1, p. 436, Table 1 (G. M. Schwab and H. Kral)	Table 5–11
vol. 1, p. 445, Fig. 3 (S. Malinowski, S. Basinski, S. Szczepanska and W. Kiewliecs)	Fig. 5–29
vol. 2, p. 1293, Table 3 (W. H. Wilmot, R. T. Barth and D. S. MacIver)	Table 5–8

Chapter 1

Solid Acids and Bases

There are many definitions of acids and bases in the literature, notably those of Arrhenius,[1] Franklin,[2] Brønsted,[3] Germann,[4] Lewis,[5] Ussanowitch,[6] Bjerrum,[7] Johnson,[8] Lux, Flood *et al.*, and Tomlinson,[9] Shatenshtein,[10] and Pearson.[11] We may understand a solid acid in general terms as a solid on which the colour of a basic indicator changes, or as a solid on which a base is chemically adsorbed. More strictly, following both the Brønsted and Lewis definitions, a solid acid shows a tendency to donate a proton or to accept an electron pair, whereas a solid base tends to accept a proton or to donate an electron pair. These definitions are adequate for an understanding of the acid-base phenomena shown by various solids, and are convenient for the clear description of solid acid and base catalysis.

In accordance with the above definitions a summarized list of solid acids and bases is given here in Tables 1–1 and 1–2. Later chapters will detail the amount, strength, and nature of acid and base centres on various solid surfaces. They will also cover the dependence of these characteristics not only upon the purity of the materials and the method of preparation, but also upon heat-treatment, compression, and irradiation.

The first group of solid acids, which includes naturally occurring clay minerals, has the longest history. Some were investigated as long ago as the turn of the century, and especially since the 1920's there have been numerous studies of their catalytic activities, although only recently have investigations commenced on zeolites. The main constituents of the first group of solid acids are silica and alumina. The very well-known solid acid catalyst synthetic silica-alumina is listed in the 4th group, which also includes the many oxide mixtures which have recently been found to display acidic properties and catalytic activity. In the 5th group are included many inorganic chemicals such as the metal oxides, sulfides, sulfates, nitrates and phosphates. Many in this group have recently been found to show characteristic selectivities as catalysts.

Of the solid bases listed in Table 1–2 special mention should perhaps be made of the alkaline earth metal oxides in the 4th group, whose basic properties and catalytic action have recently been investigated. The fact that alumina, zinc oxide and silica-alumina show both acidic and basic properties is of special significance for acid-base bifunctional catalysis. It is to be hoped that the remarkable progress made in methods of measuring acidic and basic properties (as described in the following chapters) will result in the discovery of many more solid acids and bases.

TABLE 1–1 Solid acids

1.	Natural clay minerals: kaolinite, bentonite, attapulgite, montmorillonite, clarit, fuller's earth, zeolites
2.	Mounted acids: H_2SO_4, H_3PO_4, H_3BO_3, $CH_2(COOH)_2$ mounted on silica, quartz sand, alumina or diatomaceous earth
3.	Cation exchange resins
4.	Mixtures of oxides: $SiO_2 \cdot Al_2O_3$, $B_2O_3 \cdot Al_2O_3$, $Cr_2O_3 \cdot Al_2O_3$, $MoO_3 \cdot Al_2O_3$, $ZrO_2 \cdot SiO_2$, $Ga_2O_3 \cdot SiO_2$, $BeO_2 \cdot SiO_2$, $MgO \cdot SiO_2$, $CaO \cdot SiO_2$, $SrO \cdot SiO_2$, $Y_2O_3 \cdot SiO_2$, $La_2O_3 \cdot SiO_2$, $SnO \cdot SiO_2$, $PbO \cdot SiO_2$, $MoO_3 \cdot Fe_2(MoO_4)_3$, $MgO \cdot B_2O_3$, $TiO_2 \cdot ZnO$
5.	Inorganic chemicals: ZnO, Al_2O_3, TiO_2, CeO_2, As_2O_3, V_2O_5, SiO_2, Cr_2O_3, MoO_3, ZnS, CaS, $CaSO_4$, $MnSO_4$, $NiSO_4$, $CuSO_4$, $CoSO_4$, $CdSO_4$, $SrSO_4$, $ZnSO_4$, $MgSO_4$, $FeSO_4$, $BaSO_4$, $KHSO_4$, K_2SO_4, $(NH_4)_2SO_4$, $Al_2(SO_4)_3$, $Fe_2(SO_4)_3$, $Cr_2(SO_4)_3$, $Ca(NO_3)_2$, $Bi(NO_3)_3$, $Zn(NO_3)_2$, $Fe(NO_3)_3$, $CaCO_3$, BPO_4, $FePO_4$, $CrPO_4$, $Ti_3(PO_4)_4$, $Zr_3(PO_4)_4$, $Cu_3(PO_4)_2$, $Ni_3(PO_4)_2$, $AlPO_4$, $Zn_3(PO_4)_2$, $Mg_3(PO_4)_2$, $AlCl_3$, $TiCl_3$, $CaCl_2$, $AgCl$, $CuCl$, $SnCl_2$, CaF_2, BaF_2, $AgClO_4$, $Mg_2(ClO_4)_2$
6.	Charcoal heat-treated at 300 °C

TABLE 1–2 Solid bases

1.	Mounted bases: NaOH, KOH mounted on silica or alumina; alkali metal and alkaline earth metal dispersed on silica, alumina, carbon, K_2CO_3 or in oil; NR_3, NH_3, KNH_2 on alumina; Li_2CO_3 on silica
2.	Anion exchange resins
3.	Mixtures of oxides: $SiO_2 \cdot Al_2O_3$, $SiO_2 \cdot MgO$, $SiO_2 \cdot CaO$, $SiO_2 \cdot SrO$, $SiO_2 \cdot BaO$
4.	Inorganic chemicals: BeO, MgO, CaO, SrO, BaO, SiO_2, Al_2O_3, ZnO, Na_2CO_3, K_2CO_3, $KHCO_3$, $(NH_4)_2CO_3$, $CaCO_3$, $SrCO_3$, $BaCO_3$, $KNaCO_3$, $Na_2WO_4 \cdot 2H_2O$, KCN
5.	Charcoal heat-treated at 900 °C or activated with N_2O, NH_3 or $ZnCl_2$–NH_4Cl–CO_2

REFERENCES

1. R. P. Bell, *Acids and Bases*, p. 5, Methuen, 1952.
2. E. C. Franklin, *Am. Chem. J.*, **20,** 820 (1898); **47,** 285 (1912); *J. Am. Chem. Soc.*, **27,** 820 (1905); **46,** 2137 (1924).
3. J. N. Brønsted, *Rec. Trav. Chim.*, **42,** 718 (1923); *J. Phys. Chem.*, **30,** 777 (1926); *Chem. Rev.*, **5,** 231, 284 (1928); *Z. Phys. Chem.*, **A169,** 52 (1934).
4. A. F. O. Germann, *J. Am. Chem. Soc.*, **47,** 2461 (1925).
5. G. N. Lewis, *J. Franklin Inst.*, **226,** 293 (1938); *Valency and Structures of Atoms and Molecules*, Chemical Catalog Co., 1923.
6. M. Ussanowitch, *J. Gen. Chem. USSR* (*Eng. Transl.*), **9,** 182 (1939); H. Gehlen, *Z. Phys. Chem.*, **203,** 125 (1954).
7. J. Bjerrum, *Fys. Tidssk.*, **48,** 1 (1950); B. Sansoni, *Naturwissenschaften*, **38,** 461 (1951).
8. R. E. Johnson, T. H. Norris and J. L. Huston, *J. Am. Chem. Soc.*, **73,** 3052 (1951).
9. H. Lux, *Z. Elektrochem.*, **45,** 303 (1939); H. Flood and T. Förland, *Acta Chem. Scand.*, **1,** 592, 781 (1947); J. W. Tomlinson, *The Physical Chemistry of Melts* (A symposium on molten slags and salts), p. 22, Institution of Mining and Metallurgy, 1953.
10. A. I. Shatenshtein, *Advances in Physical Organic Chemistry*, vol. 1, p. 174, Academic Press, 1963.
11. R. G. Pearson, *J. Am. Chem. Soc.*, **85,** 3533 (1963).

Chapter 2

Determination of Acidic Properties on Solid Surfaces

A complete description of acidic properties on solid surfaces requires the determination of the acid strength, and of the amount and nature (Brønsted or Lewis acid type) of the acid centres.

2.1 Acid strength

The acid strength of a solid is the ability of the surface to convert an adsorbed neutral base into its conjugate acid as described by Walling.[1] If the reaction proceeds by means of proton transfer from the surface to the adsorbate, the acid strength is expressed by the Hammett acidity function H_0[2]

$$H_0 \equiv -\log a_{H^+} f_B / f_{BH^+} \tag{2.1}$$

or

$$H_0 = pK_a + \log [B]/[BH^+] \tag{2.2}$$

where a_{H^+} is the proton activity, [B] and $[BH^+]$ are respectively the concentrations of the neutral base and its conjugate acid, and f_B and f_{BH^+} the corresponding activity coefficients. If the reaction takes place by means of electron pair transfer from the adsorbate to the surface, H_0 is expressed by

$$H_0 \equiv -\log a_A f_B / f_{AB} \tag{2.3}$$

or

$$H_0 = pK_a + \log [B]/[AB] \tag{2.4}$$

where a_A is the activity of the Lewis acid or electron pair acceptor.

2.1.1 Visual colour change method

The colour of suitable indicators adsorbed on a surface will give a measure of its acid strength: if the colour is that of the acid form of the indicator, then the value of the H_0 function of the surface is equal to or lower than the pK_a of the indicator. Lower values of H_0, of course, correspond to greater acid strength. Thus for indicators undergoing colour changes in this way, the lower the pK_a, the greater is the acid strength of the solid. For example, a solid which gives a yellow colouration with benzalacetophenone (pK_a=−5.6), but is colourless with anthraquinone (pK_a=−8.2), has an acid strength H_0 which lies between −5.6 and −8.2. A solid having $H_0 \leqq -8.2$ will change all indicators given in Table 2–1 from the basic to the acidic colours, whereas one which changed none of them would have an acid strength of H_0=+6.8 or even less (H_0 >+6.8).

In practice, the determination is made by placing about 0.2 ml of the sample in powder form into a test tube, adding 2 ml of non-polar solvent containing about 0.2 mg of indicator, and shaking briefly. Adsorption, if it occurs at all, proceeds very rapidly, and the change in colour between basic and acidic forms of the indicator is most striking. Table 2–1 lists those basic indicators which have been used. The solvents used are benzene, isooctane, decalin and cyclohexane. H_R indicators (a series of allylalcohols which react with acids to form carbonium ions)

TABLE 2–1 Basic indicators used for the measurement of acid strength†1

Indicators	Colour		pK_a	$[H_2SO_4]$†2 (%)
	Base-form	Acid-form		
Neutral red	yellow	red	+6.8	8×10^{-8}
Methyl red	yellow	red	+4.8	—
Phenylazonaphthylamine	yellow	red	+4.0	5×10^{-5}
p-Dimethylaminoazobenzene (Dimethyl yellow or Butter yellow)	yellow	red	+3.3	3×10^{-4}
2-Amino-5-azotoluene	yellow	red	+2.0	5×10^{-3}
Benzeneazodiphenylamine	yellow	purple	+1.5	2×10^{-2}
4-Dimethylaminoazo-1-naphthalene	yellow	red	+1.2	3×10^{-2}
Crystal violet	blue	yellow	+0.8	0.1
p-Nitrobenzeneazo-(*p*′-nitro)diphenylamine	orange	purple	+0.43	—
Dicinnamalacetone	yellow	red	−3.0	48
Benzalacetophenone	colourless	yellow	−5.6	71
Anthraquinone	colourless	yellow	−8.2	90

†1 *o*-Chloroaniline (pK_a = −0.17) and *p*-chloro-*o*-nitroaniline (pK_a = −0.91) are not suitable, since the acid-forms are colourless.

†2 wt% of H_2SO_4 in sulfuric acid solution which has the acid strength corresponding to the respective pK_a.

are also available for acid strength determinations in certain specialized applications.[3]

The above indicator method was originally reported by Walling,[1] Weil-Malherbe and Weiss,[4] and Ikebe *et al.*[5] and has been used extensively by many workers. The acid strengths of some solids determined by this method are given in Table 2–2.[6-9] Although subject to the limitations mentioned in the following section, this method readily enables a determination of relative acid strengths. The usual colour change, however, is difficult to observe when the sample is itself coloured or dark, but the difficulty can be minimized by mixing a white substance of known acidity with the sample (see 2.2.2) or by employing the spectrophotometric method outlined in the following section.

We may note here the convenience and desirability of measuring the number of acid sites by amine titration (see 2.2.1) immediately after a determination of acid strength by this method.

TABLE 2–2 Acid strength of some solids

Solid acids	H_0	Reference No.
Original kaolinite	$-3.0 \sim -5.6$	6
Hydrogen kaolinite	$-5.6 \sim -8.2$	6
Original montmorillonite	$+1.5 \sim -3.0$	6
Hydrogen montmorillonite	$-5.6 \sim -8.2$	6
$SiO_2 \cdot Al_2O_3$	< -8.2	6
$Al_2O_3 \cdot B_2O_3$	< -8.2	6
$SiO_2 \cdot MgO$	$+1.5 \sim -3.0$	6
1.0 mmol/g H_3BO_3/SiO_2	$+1.5 \sim -3.0$	6
1.0 mmol/g H_3PO_4/SiO_2	$-5.6 \sim -8.2$	6
1.0 mmol/g H_2SO_4/SiO_2	< -8.2	6
$NiSO_4 \cdot xH_2O$ heat-treated at 350°C	$+6.8 \sim -3.0$	7
$NiSO_4 \cdot xH_2O$ heat-treated at 460°C	$+6.8 \sim +1.5$	7
ZnS heat-treated at 300°C	$+6.8 \sim +4.0$	8
ZnS heat-treated at 500°C	$+6.8 \sim +3.3$	8
ZnO heat-treated at 300°C	$+6.8 \sim +3.3$	9
TiO_2 heat-treated at 400°C	$+6.8 \sim +1.5$	9

2.1.2 Spectrophotometric method

In view of the uncertainties inherent in Hammett indicator determinations of surface acidity by visual means, a study has been made of the characteristic absorption spectra of dyes adsorbed on several silica-alumina catalysts and on silica gel.[10,11] Hammett indicators shown in Table 2–3 were adsorbed from isooctane or benzene onto thin plates of optically transparent catalysts. Those indicators whose pK_a values are equal to or greater than the characteristic H_0 of an acid surface are adsorbed in their acid form. Fig. 2–1 shows the absorption spectra for phenylazonaphthylamine (pK_a=+4.0) adsorbed on silica-alumina (12% Al_2O_3)[10]. Curves a and b are the spectra of the basic and acidic forms of the indicators as obtained respectively in isooctane and in ethanolic HCl acid solution. In this case curve c clearly reveals that the indicator is adsorbed exclusively in its acid form.

Employing various of the basic indicators in Table 2–3, Dzisko *et al.* were able by this method to determine the acid strength of metal oxide mixtures.[12] The order of the acid strength was found to be $SiO_2 \cdot Al_2O_3 > ZrO_2 \cdot SiO_2 \sim Ga_2O_3 \cdot SiO_2 > BeO \cdot SiO_2 \sim MgO \cdot SiO_2 > Y_2O_3 \cdot SiO_2 > La_2O_3 \cdot SiO_2 > SnO \cdot SiO_2 \sim PbO \cdot SiO_2$.

TABLE 2–3 Basic indicators used for spectrophotometric determination of acid strength

Indicators	pK_a	pK_{R^+}
Phenylazonaphthylamine	+4.0	
p-Dimethylaminoazobenzene	+3.3	
Aminoazobenzene	+2.8	
Benzeneazodiphenylamine	+1.5	
p-Nitroaniline	+1.1	
o-Nitroaniline	−0.2	
p-Nitrodiphenylamine	−2.4	
2,4-Dichloro-6-nitroaniline	−3.2	
p-Nitroazobenzene	−3.3	
2,4-Dinitroaniline	−4.4	
Benzalacetophenone	−5.6	
p-Benzoyldiphenyl	−6.2	
Anthraquinone	−8.1	
2,4,6-Trinitroaniline	−9.3	
3-Chloro-2,4,6-trinitroaniline	−9.7	
p-Nitrotoluene	−10.5	
Nitrobenzene	−11.4	
2,4-Dinitrotoluene	−12.8	
Triphenylcarbinol		−6.6
Diphenylcarbinol		−13.3
2,4,6-Trimethylbenzyl alcohol		−17.4

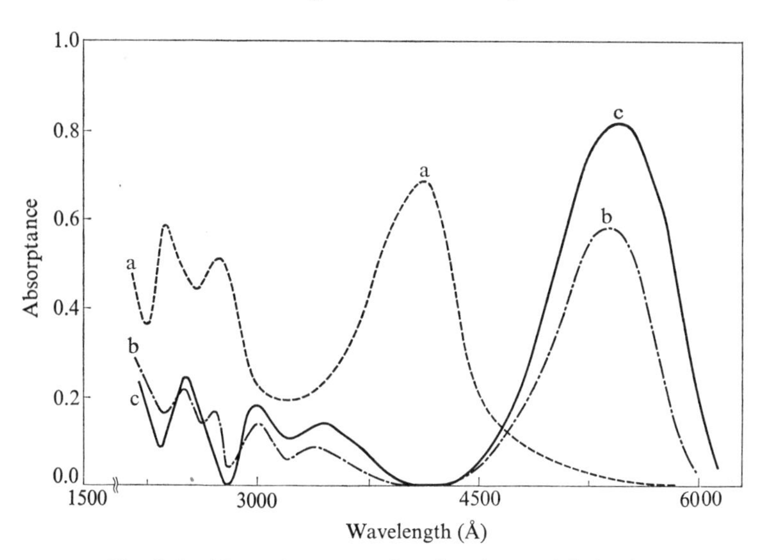

Fig. 2–1 Absorption spectra for phenylazonaphthylamine
a: in isooctane solution, b: in ethanolic HCl, c: adsorbed on silica-alumina
N.B.: Absorptance—often referred to as "absorbance"

Kobayashi investigated the absorption spectra of dimethyl yellow, methyl red, and bromophenol blue adsorbed on silica-alumina which had been partially covered with the organic base, *n*-butylamine, in a non-polar solvent and developed a method for determining both acid strength and total amount of acid.[13] Assuming that both *n*-butylamine (B) and indicator (I) molecules are in equilibrium with the acid sites ($\overline{HS}$) and the basic sites ($\overline{S}$), he derived the following equation:

$$Cx^2 - K_1(A-C)x - K_1A/K_2 = 0 \tag{2.5}$$

where C is the concentration of amine on the surface, x the ratio of the acidic colour to the basic in the indicator spectrum, and A the concentration of acid sites. K_1 and K_2 are the equilibrium constants given by $[I\overline{HS}][\overline{S}]/[\overline{HS}][I\overline{S}]$ and $[B\overline{HS}][I\overline{S}]/[I\overline{HS}][B\overline{S}]$ respectively. K_1, which is by definition the acid strength of the surface, as well as A and K_2, can be obtained from experimental data on the C–x relationship.

On the basis of these spectrographic studies Kobayashi also discussed the validity of taking the H_0 function as a measure of surface acidity.[14] By varying the concentration C of the adsorbed organic base (*n*-butylamine as above, or pyridine) and measuring the corresponding ratios $[B]/[BH^+]$ of the basic to the acidic form of the indicator, values of H_0 can be obtained from Eq. 2.2 for each value of C. At any given C,

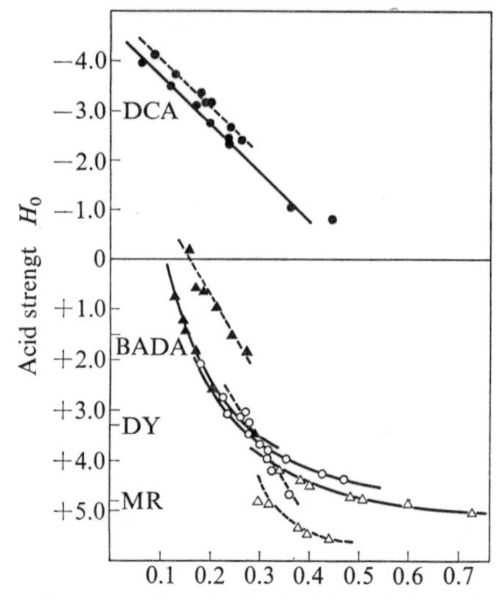

Fig. 2–2 Values of H_0 for $SiO_2 \cdot Al_2O_3$ at different concentrations of adsorbed organic base
(—●—) dicinnamalacetone (DCA)/pyridine (Py),
(--●--) DCA/*n*-butylamine (Bta),
(—▲—) benzeneazodiphenylamine (BADA)/Py,
(--▲--) BADA/Bta, (—○—) dimethyl yellow (DY)/Py,
(--○--) DY/Bta, (—△—) methyl red (MR)/Py,
(--△--) MR/Bta

the values obtained for H_0 should, of course, be independent of the type of indicator used. In fact the plots of H_0 against C lie on several distinct straight lines or curves as shown in Fig. 2–2. The discrepancy is considered to be due mainly to the difference in molecular structure, colour change mechanism, and the amount and activity of the indicator used. There is also the effect of discrepancies between the well-established pK_a values for aqueous solutions, and the undetermined pK_a values for this type of system. It is therefore important to select indicators which come from analogous chemical groups and to apply proper corrections to pK_a values.

2.1.3 Gaseous base adsorption method

When gaseous bases are adsorbed on acid sites, a base adsorbed on a strong acid site is more stable than one adsorbed on a weak acid site, and is more difficult to desorb. As elevated temperatures stimulate evacuation of the adsorbed bases from acid sites, those at weaker sites will be evacuated preferentially. Thus, the proportion of adsorbed base evacuated at various temperatures can give a measure of acid strength. Using various $HF–Al_2O_3$ catalysts on which ammonia had been adsorbed at 10 mmHg at 175 °C for 30 min, Webb determined the fraction of ammonia remaining on the surface after evacuation at various temperatures between 175 and 500 °C.[15] As shown in Fig. 2–3, the fraction of ammonia remaining was found to be larger when the catalyst contained a higher proportion of hydrogen fluoride. That is, the catalyst acid strength increases with increasing hydrogen fluoride content. This method also gives results in agreement with the visual colour change method in the case of silica gel and silica-alumina, for the comparative ease with which ammonia may be desorbed from silica gel indicates a weak acid strength, and the difficulty in the case of silica-alumina a correspondingly greater strength.[16]

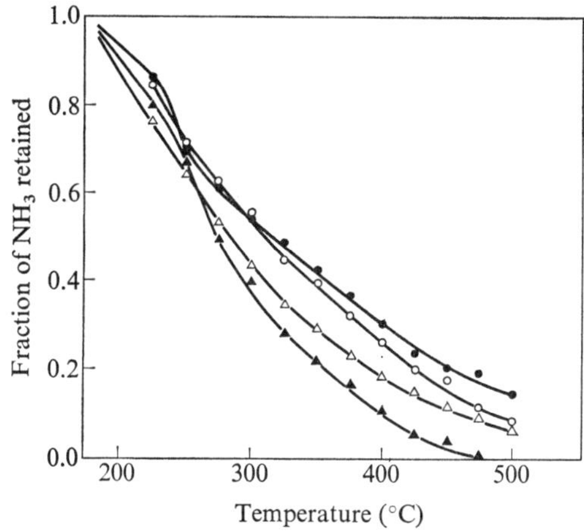

Fig. 2–3 Fraction of ammonia retained on catalyst surface *vs.* evacuation temperature for various percentages HF content
(▲) 0.0% HF content, (△) 0.65%,
(○) 3.23%, (●) 6.46%

The differential thermal analysis (DTA) method gives a convenient means of differentiating between acid strengths. Bremer and Steinberg,[17] while heating $MgO \cdot SiO_2$ catalysts which had been coated with pyridine, noted that desorption as displayed on the DTA diagram was affected by the preheating treatment given to the catalyst. Preheating at high temperature gave a pyridine desorption peak at a lower temperature, with progressively higher pyridine desorption peak temperatures as the preheating temperature was reduced. The desorption peak at lower temperature can be associated with pyridine desorption from weaker acid sites, and conversely the peaks at higher temperatures indicate stronger acid sites. In 2.2.4 another DTA method is outlined, which yields both acid amount and acid strength.

The heat of adsorption of various bases is also clearly a measure of the acid strength on the solid surface. The heat of adsorption ΔH of trimethylamine for two kinds of cracking catalysts has been measured by calorimetric methods. ΔH was found to take the values -33 ± 6 kcal/mol and -38 ± 6 kcal/mol.[18] These values are quite near those for the acid-base reaction $HCl(g) + NH_3(g) = NH_4Cl(s)$, $\Delta H = -42$ kcal/mol, so that the acid strength is considered to be fairly high, in agreement with results obtained by other methods. Zettlemoyer and Chessick[19] calculated the energy distribution of the acid sites from the relationship of the differential heat of adsorption with the amount of butylamine which had been adsorbed. The measurement of the heat of chemical adsorption of piperidine vapour has been attempted by Stone and Rase.[20] The heat of adsorption of ammonia on silica-alumina has been determined by both gravimetric and calorimetric methods,[21-23] while those of hydrogen sulfide on alumina[24] and pyridine and benzene on silica-alumina[25] have been measured by means of gas analysis and gas chromatography respectively.

An example of observations on heat of adsorption is given in Fig. 2–4, where the differential heat for ammonia adsorption on silica-

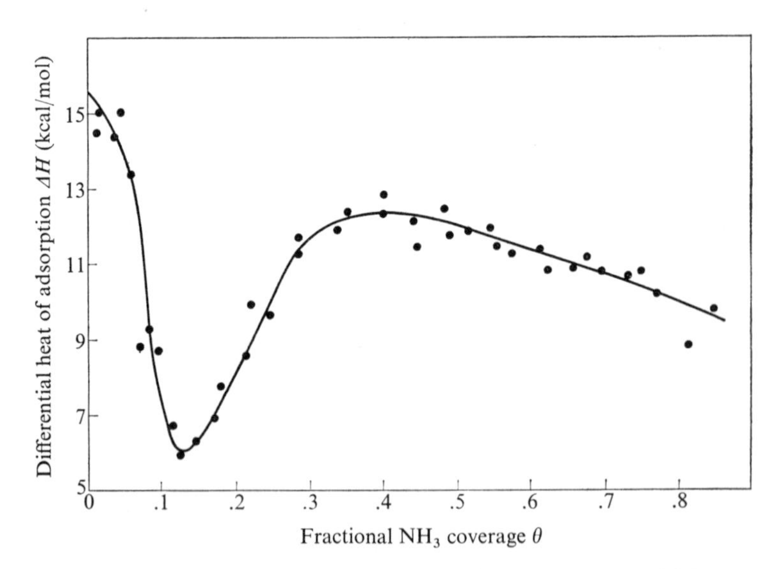

Fig. 2–4 Differential heat of adsorption ΔH for NH_3 on $SiO_2 \cdot Al_2O_3$ *vs.* fractional NH_3 coverage θ

alumina is plotted against θ, the proportional coverage of ammonia.[26] The figure shows an initial value for adsorption heat of between 9 and 15 kcal/mol, with a minimum at $\theta \cong 0.1$, and a subsequent sharp increase. At high θ, the interactions between adsorbates are considered to be fairly strong. Hsieh also discusses the effect of cracking activity upon the energy of the acid sites in this case.[21]

2.1.4 Other methods

Pines and Haag have estimated the acid strength of alumina catalysts from the respective rates of the isomerization of cyclohexane and dimethyl-1-butene and the dehydration of 1-butanol.[26] Shiba *et al.* suggest that the equilibrium constant of ammonia adsorption on a catalyst surface can give a measure of its acid strength.[27] This equilibrium constant is obtained by an analysis of the data derived from the reversible ammonia poisoning which occurs during the cracking of cumene. Certain other methods are described in the following section on the determination of the amount of acid.

2.2 Amount of acid

The amount of acid on a solid is usually expressed as the number or mmol of acid sites per unit weight or per unit surface area of the solid, and is obtained by measuring the amount of a base which reacts with the solid acid using one or other of the several methods described below. It is also sometimes loosely called "acidity".

2.2.1 Amine titration method

This method was reported first by Tamele[28] and is based on Johnson's experiment. It consists of titrating a solid acid suspended in benzene with *n*-butylamine, using *p*-dimethylaminoazobenzene as an indicator. The yellow basic form of the indicator changes to its red acidic form when adsorbed on the solid acid. Thus, the titres of *n*-butylamine required to restore the yellow colour give a measure of the number of acid sites on the surface. Since the indicator which has been adsorbed on the acid sites must be replaced by the titrating base, the basicity of the latter should be the stronger of the two.

The amount of acid indicated in this way by using *p*-dimethylaminoazobenzene ($pK_a = +3.3$) is actually the amount of those sites having acid strengths with $H_0 \leqq +3.3$. The use of various indicators with different pK_a values (see Table 2–1) enables a determination of the amount of acid *at various acid strengths* by amine titration. This method gives the sum of the amounts of both Brønsted and Lewis acid, since both electron pair acceptors and proton donors on the surface will react with either the electron pair (—N̈=) of the indicator or that of the amine (≡N:) to form a coordination bond.

$$C_6H_5\text{—N=N—}C_6H_4\text{—N(CH}_3)_2 + A$$

(yellow form of indicator) (solid acid)

$$\rightleftharpoons C_6H_5\text{—N(A)—N=}C_6H_4\text{=N}^+(CH_3)_2$$

(red form of indicator)

$$C_6H_5\text{—N(A)—N=}C_6H_4\text{=N}^+(CH_3)_2 + C_4H_9NH_2$$

$$\rightleftharpoons \text{yellow form of indicator} + C_4H_9NH_2\cdot A$$

(*n*-butylamine adsorbed on solid acid)

Let us consider a practical example of the calculation of acid amount. In the titration of 0.1 N *n*-butylamine against 0.5 g of nickel sulfate powder (100～200 mesh) heat-treated at 300 °C, 0.55 ml of titre was necessary with dimethyl yellow as the indicator.[29] Thus, the amount of acid is calculated to be 0.11 mmol/g or 0.11 meq/g. Since the surface area of the solid is known to be 11 m^2/g, we can also express this result as 0.010 mmol/m^2.

The convenience of the above method has led to its extensive use by many workers since Tamele. Johnson showed that carbon tetrachloride and isooctane can also be used as solvents, and benzylamine as the titrating base.[30] He showed that, unlike the case of nickel sulfate, where titration may be practically completed in 15～30 min, it is necessary to take 2 or 3 days for the titration with silica-alumina in order to minimize the adsorption of amine on the non-acidic portions of the surface. A drop of 0.05 N trichloroacetic acid in benzene was added to ensure that there was no appreciable excess of amine on the solid at the end point. However, a drop by drop amine titration extending over 2 or 3 days is troublesome. This led Benesi to modify the titration technique so that the indicators could be added to portions of the catalyst in suspension after the catalyst sample had reached equilibrium with *n*-butylamine, the end point being determined by a series of successive approximations.[31] The modified procedure has several advantages;

1) end points for titration of a particular catalyst can be determined by means of as many as 10 different indicators with little more effort than that for a single indicator, 2) *n*-butylamine can be brought to equilibrium with the catalyst sample without the introduction of traces of moisture in the course of titration, and 3) equilibrium is attained with greater rapidity when the indicator is added to the *n*-butylamine rather than by adding the amine until the strongly adsorbed acid form of the indicator has been displaced.

The effects upon the acid site determination of varying both the amount of indicator (dimethyl yellow) and the length of time taken for the titration of *n*-butylamine against silica-alumina by Johnson's method are shown in Fig. 2–5 A and B.[32] The amount of acid as measured is almost independent of the amount of indicator added beyond 0.2 ml and changes very little for titration times longer than 50 h. Fig. 2–6 A shows the effect of powder size upon acid amount. Even with a titration time of only 2 h, provided the sample is a powder of 100 mesh or finer, constant values for the amount of acid are obtained. Fig. 2–6 B shows the effect of moisture on acid amount, where the amount with $H_0 \leqq -3$ is found to decrease markedly when the sample is exposed in an atmos-

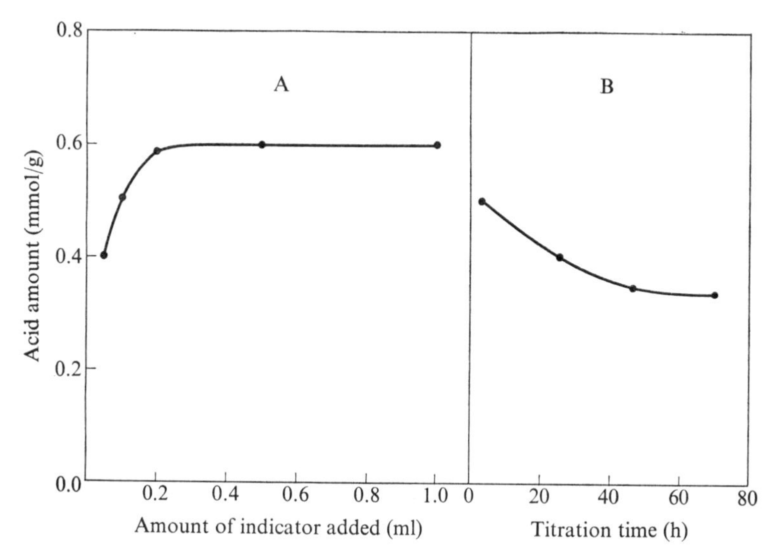

Fig. 2–5 A Effect of added indicator volume on measured acid amount
Sample: 0.5 g $SiO_2 \cdot Al_2O_3$ (<100 mesh) in 10 ml benzene. Indicator: dimethyl yellow (1% benzene solution).
B Effect of titration time on measured acid amount
Sample: 0.5 g $SiO_2 \cdot Al_2O_3$ (<100 mesh) in 10 ml benzene. Added volume of indicator: 0.3 ml benzeneazodiphenylamine (1% benzene solution).

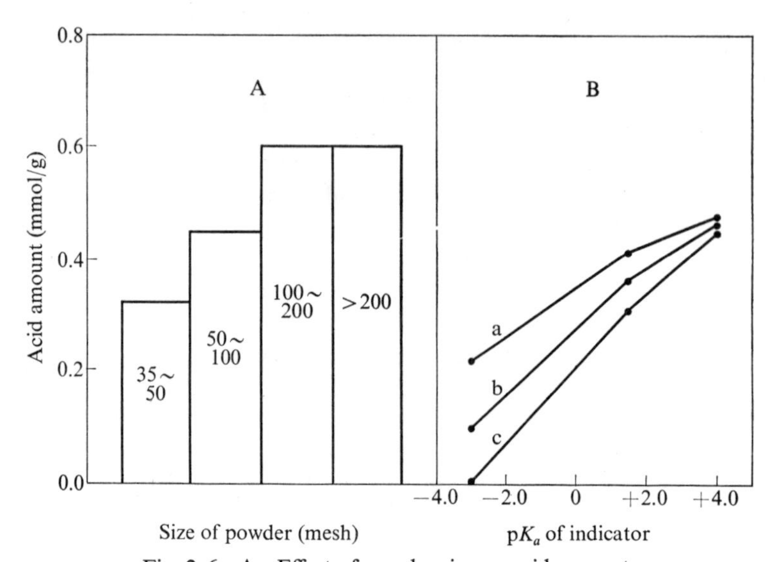

Fig. 2–6 A Effect of powder size on acid amount
Sample: 0.5 g $SiO_2 \cdot Al_2O_3$ in 10 ml benzene. Added volume of indicator: 0.3 ml dimethyl yellow (1% benzene solution). Titration time: 2 h.
B Effect of moisture on acid amount
Sample: 0.5 g $SiO_2 \cdot Al_2O_3$ (<100 mesh) in 10 ml benzene. Added volume of indicators: 0.3 ml each of several pK_a values (1% benzene solution). Titration time: 2 h. a: dried catalyst, b: catalyst left in 90% humidity at 20 °C for 5 min, c: catalyst left in 90% humidity at 20 °C for 10 min.

phere of 90% humidity for 5 min, almost all the acid sites being poisoned after a 10 min exposure.[32] However, there is a net increase in the amount of acid with H_0 lying between −3 and +1.5 (that is the difference between the amounts for $H_0 \leqq -3$ and $H_0 \leqq +1.5$). This seems to indicate that the adsorption of water molecules changes a number of the stronger acid sites into weaker ones. (A similar phenomenon is observed when alumina and silica-alumina are gradually poisoned with different amounts of sodium hydroxide.) Moisture does not appear to have much effect on the total amount of acid sites at $H_0 \leqq +4.0$. Since the pK_a value of water is −1.7, poisoning consists of the adsorption of water molecules upon acid sites with H_0 lower than −1.7.

In 2.1.2 a method was described for determining both acid amount and acid strength by observing the absorption spectra of acidic and basic forms of Hammett indicators adsorbed on the solid surface. Spectrographic methods can be extended to include coloured samples, as described in the next section, provided only that the catalyst is optically transparent.

2.2.2 Amine titration method for coloured samples

The amine titration method described in the foregoing section is obviously limited to white or light-coloured surfaces. Titrations of dark coloured solids can be carried out, however, by adding a small known amount of a white solid acid.[30] The end-point of the titration is taken when the colour change is observed on the white solid, and a correction is made for the amount of butylamine used for the added white material. Using this method, both acid amount and acid strength have been measured for titanium trichloride by employing silica-alumina as the white material.[33] Since titanium trichloride sinks to the bottom of the titration flask immediately after shaking the mixture in benzene, whereas the silica-alumina remains in suspension for a while, the colour change is not very difficult to detect. The sharpest colour change is observed for a mixture with the proportion of 0.02 ~ 0.05 g titanium trichloride to *ca.* 0.2 g silica-alumina. It becomes difficult to observe the colour change with more than 0.15 g of titanium trichloride even if the amount of silica-alumina is increased in the same proportion as that mentioned above. Naturally the accuracy suffers if the relative proportion of silica-alumina is increased. The anomalously large values for acid amount found for titanium trichloride by this method are cited in 4.5.1. A similar method using alumina as the white material was employed to measure the acid amount of dark green chromic oxide.[34] A chromic oxide sample, heat-treated at 500 °C for 4 h was found to have the following acid amounts: 0.09 mmol/g at $H_0 \leqq +3.3$ and at $H_0 \leqq -3$, and 0.05 mmol/g at $H_0 \leqq -8.2$.

The acid amount of coloured solids can be determined also by calorimetric titration, using *n*-butylamine, ethylacetate or dioxane as the titrating base. Calorimetric determinations were first carried out by Trambouze *et al.*[35] and developed by Topchieva *et al.*[36] and Tanabe and Yamaguchi.[37] The detailed procedure adopted with *n*-butylamine is described below.

Benzene (~ 100 ml) and 3 ~ 5 g of solid materials were put into a 500 ml Dewar flask equipped with a Beckmann thermometer, glass stirrer and a microburette. A thick cork was used to improve thermal insulation. The contents of the vessel were stirred and the slight rise in temperature due to the mechanical heat generation were read every 2 min as shown by the solid circles on line "M" in Fig. 2–7.[37] Then 0.5 ml of *n*-butylamine were added from a microburette, and the temperature rise due to the heat of the acid-base reaction was read every 30 s (later every 60 s) as shown by the open circles in Fig. 2–7. After the rate of temperature increase reduced to the constant value due to stirring, as

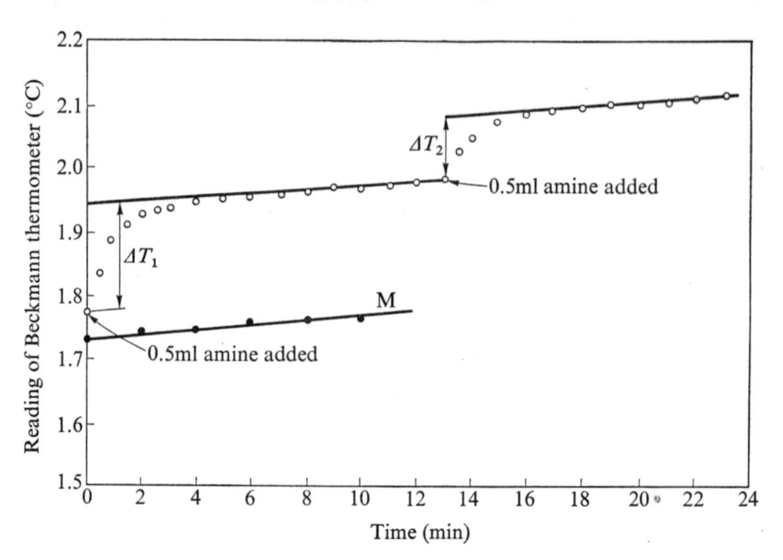

Fig. 2–7 Temperature rise due to the heat of reaction of 3.9 g silica-alumina with 0.5 ml 1.01 N *n*-butylamine
Line M shows the temperature rise where no amine is added

indicated by a temperature–time relation parallel with line M, another 0.5 ml (later increased to 1.0 ml) of the amine was added and the temperature rise was recorded similarly. This procedure was repeated until no further temperature increase due to the reaction could be brought about by further addition of amine. The temperature increase due to the heat of acid-base reaction is given by the difference ΔT_i between parallel lines before and after each addition of amine.

The integrated increase in temperature, $\Delta T_n = \Sigma \Delta T_i$ for n additions of amine was plotted against the cumulative volume of amine titre as shown in Fig. 2–8. At the end point of the titration the curve flattens out, indicating that for a certain n, $\Delta T_n = \Delta T_{n+1}$. The acid amount is obtained from the total amount of amine which corresponds to this end point.

Fig. 2–8 shows that 3 ml of 1.01 N *n*-butylamine are required to neutralize 3.9 g of silica-alumina. Calculation yields an acid amount of 0.78 mmol/g. This is a higher value than that given by the amine titration method using methyl red ($pK_a = +4.8$), which was 0.55 mmol/g. This indicates the existence of acid sites with acid strength less than that for $H_0 = +4.8$, and shows that these weaker acid sites are susceptible to measurement by the calorimetric titration method when the comparatively strong base *n*-butylamine ($pK_a = +10.2$) is used. Fig. 2–8

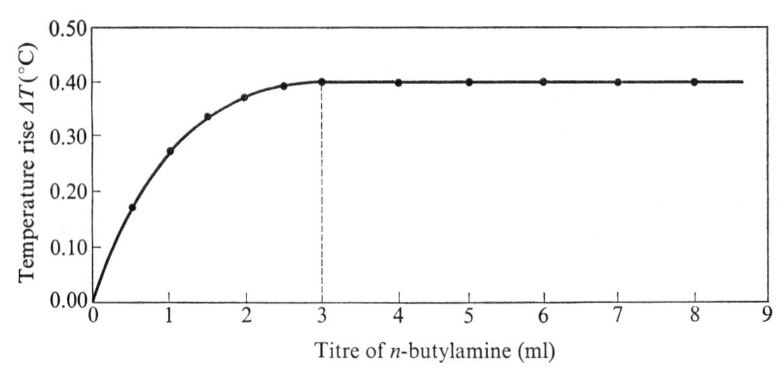

Fig. 2–8 Calorimetric titration curve: 3.9 g silica-alumina with 1.01 N *n*-butylamine (*cf.* Fig. 2–7)

also shows that the rate of temperature rise is gradually reduced as the total amount of amine titre is increased. This seems to indicate that the acid sites are energetically heterogeneous, as suggested by Topchieva *et al.* The slow reaction of amine with the weaker acid sites combined with a reduction in the rate of diffusion of amine molecules into the micro-pores of the solid catalyst may be responsible for the slower rate of temperature rise.[36)]

2.2.3 Titration in aqueous solutions

This method gives a measure of the amount of acid sites by the extent to which they neutralize an aqueous solution of potassium hydroxide as revealed by subsequent titration with hydrochloric acid using phenolphthalein as an indicator.[15,38)] This method is unreliable, however, in that 1) a strong alkali may react with some parts of the surface other than the acid sites, 2) a water molecule may react with an anhydrous substance or a Lewis acid and so change the acidic properties of the surface and 3) a solid with acid strength greater than that of an oxonium ion would be reduced to that of the oxonium ions by water molecules acting as a strong base. The method cannot, of course, be applied to solid acids which are soluble in water. Some workers use this method for the determination of the amount of a Brønsted acid, a technique described later in detail in 2.3.1.

2.2.4 Adsorption and desorption of gaseous bases

The amount of a base which a solid acid can adsorb chemically from the gaseous phase is a measure of the amount of acid on its surface. When a solid sample is put in a quartz spring balance and evacuated, the vapour of an organic base such as pyridine or trimethylamine may be introduced for adsorption. When prolonged subsequent evacuation produces no further decrease in sample weight, then the base which is retained upon the sample is understood to be chemically adsorbed.[18] In the method used by Mills *et al.*[39] nitrogen gas containing quinoline vapour is introduced to a sample within a tube, and after adsorption has taken place pure nitrogen gas is passed over the sample to remove any base that has been merely physically adsorbed, the weight increase due to chemically adsorbed base being measured by a chemical balance connected directly to the adsorption tube.

The bases used as adsorbates so far include quinoline,[39,40] pyridine,[18,41,42] piperidine,[20] trimethylamine,[18,20,43] *n*-butylamine,[19,42] pyrrole[43] and ammonia.[23,41–44] The various bases are adsorbed to different extents even by the same catalyst. Uchida and Temma measured the amounts of several bases chemically adsorbed on silica-alumina at various evacuation temperatures.[42] The considerable differences which occur at low evacuation temperatures are clearly shown in Fig. 2–9. This effect is probably due to differences in the basicity of the

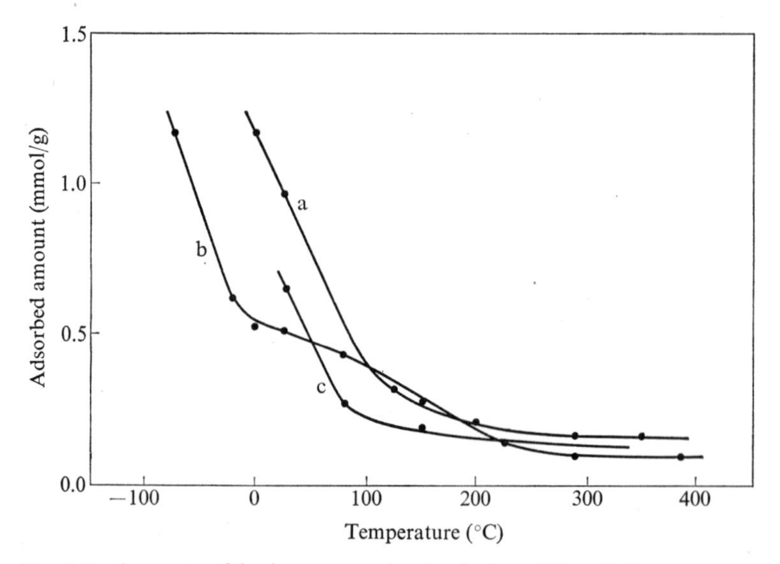

Fig. 2–9 Amounts of basic vapours chemisorbed on $SiO_2 \cdot Al_2O_3$ *vs.* evacuation temperature
a: *n*-butylamine, b: NH_3, c: pyridine

adsorbates and in the extent to which they diffuse into the interior of the catalyst micro-pores.

Murakami and Shiba showed that the adsorption isotherm of

ammonia on alumina can be interpreted in terms of two kinds of adsorption, i.e. an adsorption at low pressure which decreases in the presence of alkali, and a second adsorption which is uninfluenced by alkali.[44] The former would be the adsorption upon acid sites. Kubokawa obtained both the amount of acid sites and their acid strength for silica-alumina and the metal sulfates from the relationship between the activation energy of ammonia desorption and the amount of ammonia remaining on the surface.[45]

The differential thermal analysis method is also available for the estimation of the acid amount together with the acid strength of a solid. By DTA and thermal gravimetric analysis (TGA) of silica-alumina on which pyridine, *n*-butylamine, or acetone had been adsorbed, Shirasaki *et al.* obtained values for the amount x of the base retained on the solid (from the TGA curve), and for the corresponding amount of heat absorbed (from the area S defined by the DTA curve), see Fig. 2–10.[46]

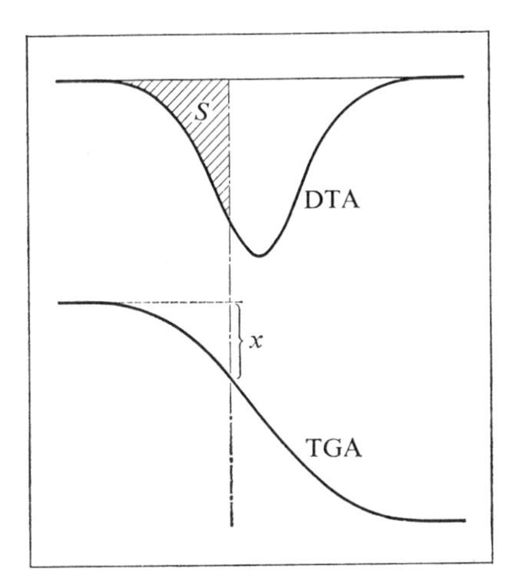

Fig. 2–10 Schematic DTA and TGA curves
See text for significance of S and x

From the curve of S against x they calculated dS/dx. Plotting x against dS/dx gives the acid amounts at the various acid strengths (or the heat required for desorption of the base).

Methods utilizing adsorption and desorption of the gaseous phase have the advantage that the acid amount for a solid at high temperatures (several hundred degrees centigrade), or under its actual working conditions as a catalyst, can be determined. The methods also apply even to coloured samples. They suffer from the disadvantage that it is

TABLE 2–4 Methods for measuring acid amount and/or strength by using gaseous bases

Method(s)	Measurement(s)	Solid acid(s)	Gaseous base(s)	Ref. No.
Gravimetric	Acid amount	$SiO_2 \cdot Al_2O_3$ $SiO_2 \cdot ZrO_2$ $SiO_2 \cdot MgO$	Pyridine Quinoline	39
		$SiO_2 \cdot Al_2O_3$ Al_2O_3	Pyridine NH_3, *n*-Butylamine	42
	Amount and strength	$SiO_2 \cdot Al_2O_3$ Al_2O_3	NH_3	23
	Acid strength and adsorption entropy	Zeolite Y (H–Y)	NH_3	47
Gravimetric-Calorimetric	Amount and strength	$SiO_2 \cdot Al_2O_3$	Pyridine Trimethyl-amine	18
		$SiO_2 \cdot Al_2O_3$ Kaolin	*n*-Butyl-amine	19
Calorimetric	Acid strength	$SiO_2 \cdot Al_2O_3$ SiO_2, Al_2O_3 Zeolite X (Ca–X, Na–X)	NH_3	48
		$SiO_2 \cdot MgO$ Al_2O_3	NH_3	22
	Amount and strength	$SiO_2 \cdot Al_2O_3$	NH_3	21
Volumetric	Acid amount	Al_2O_3	Pyridine NH_3	41
		Al_2O_3	NH_3	44
Volumetric-Gravimetric	Amount and strength	$SiO_2 \cdot Al_2O_3$	Trimethyl-amine Pyrrole, NH_3	43
DTA	Acid strength	$SiO_2 \cdot Al_2O_3$	Piperidine	20
DTA/TGA	Amount and strength	$SiO_2 \cdot Al_2O_3$	*n*-Butylamine Pyridine Acetone Methyl-ethylketone	46
NH_3 Desorption	Acid amount	Al_2O_3	NH_3	15
		$SiO_2 \cdot Al_2O_3$ SiO_2, Al_2O_3 Ni/Al_2O_3 Ni/Kieselguhr	NH_3	49
NH_3 Desorption rate	Amount and strength	$SiO_2 \cdot Al_2O_3$ Al_2O_3 H_2SO_4/SiO_2	NH_3	45
NH_3 Flash desorption	Acid strength	Al_2O_3	NH_3	50
Gas chromatography	Amount and strength	$SiO_2 \cdot Al_2O_3$	Pyridine Benzene	25
Gas analysis	Amount and strength	Al_2O_3	H_2S	24
Thermal conductivity	Acid amount	Zeolite Y $SiO_2 \cdot Al_2O_3$	Quinoline	51

difficult to distinguish between chemical and physical adsorption and to differentiate between the amounts of acid at various acid strengths. Table 2–4 summarizes the various methods for measuring acid amount or acid strength using gaseous bases.

2.2.5 Other methods

A comparison between the rate constants of certain reactions using solid acid catalysts and those catalyzed by the usual homogeneous acid catalysts can sometimes yield an estimate of surface acid amount. Suitable reactions include the conversion of sucrose[40,52] and the esterification of phthallic acid.[53]

In the decomposition of cetane, when varying amounts of lithium hydroxide are added to the catalyst, the amount of active acid sites can be estimated from the relationship between the amount of lithium hydroxide and the percentage of decomposition in the reaction.[54]

Both the amount of carbon dioxide evolved in the reaction of a solid acid with sodium bicarbonate[55] and the amount of hydrochloric acid gas evolved in the reaction between a monohydrated metal sulfate and solid sodium chloride[56] can give a measure of the acid amount for these solids.

2.3 Brønsted and Lewis acid sites

Each of the methods for determining the strength and amount of acid described in the foregoing sections (2.1 and 2.2) do not distinguish between Brønsted acid sites and Lewis acid sites. The acid amount which is measured is the sum of the amounts of Brønsted acid and Lewis acid (which we shall sometimes abbreviate conveniently to B acid and L acid) at a certain acid strength. In order to elucidate the catalytic actions of solid acids it will often be necessary to distinguish between B acids and L acids.

2.3.1 The measurement of Brønsted acid alone

The number of B acid sites on a solid surface may be derived from the number of free protons in aqueous solution arising from the exchange of a proton (or a hydrogen atom which can be liberated as a proton) with the cation of a salt in aqueous solution at the surface of the solid. For this purpose 5% sodium chloride solution[57] or 0.1 N ammonium acetate solution[58] are used as the salt. This proton exchange approach has been used by Mähl,[59] Trambouze *et al.*,[60] Holm *et al.*[61] and

Danforth.[62] By using ammonium acetate solution, Holm *et al.* were able not only to determine the number of B acid sites on silica-alumina, but also their acid strength $K=[H^+][\overline{NH_4}]/[\overline{H}][NH_4^+]$, where $\overline{H}$ and $\overline{NH_4}$ denote hydrogen and ammonia on the surface. A comparison between the amount of B acid observed by this method and the total amount of B plus L acids as measured by amine titration, and the correlation between the number of B sites and their activity in propylene polymerization is shown in 5.1.3. The various methods given in 2.2.5 also determine only the amount of B acid, but those methods using aqueous solutions seem unreliable for the reasons given in 2.2.3.

Malinowski and Szczepanska measured the amount of B acid by potentiometric titration of a solid acid in anhydrous picoline with a 0.1 N solution of sodium ethoxide.[63] Their results obtained with Na/SiO_2 catalyst are shown in Fig. 2–11.

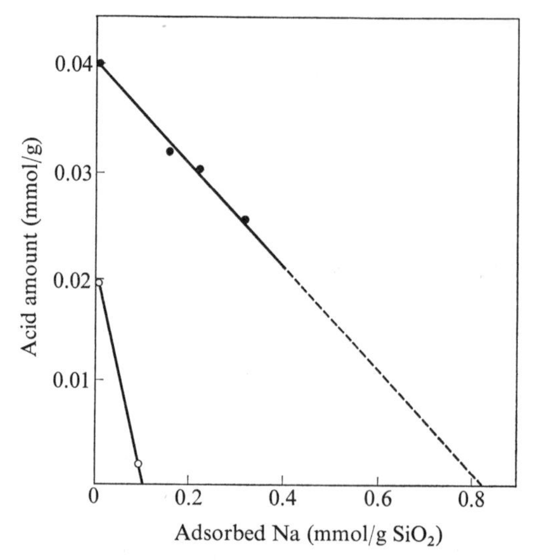

Fig. 2–11 Brønsted acidity on Na/SiO_2 catalyst (●) catalyst dried at 120 °C, (○) catalyst calcined at 1,000 °C

2.3.2 The measurement of Lewis acid alone

The first method we discuss makes use of certain reagents which react with L acids but not with B acids. Okuda and Tachibana observed that the cation radical $[NH_2C_6H_4NH_2\cdot]^+$ (Würster salt) of the reagent *p*-phenylenediamine ($NH_2C_6H_4NH_2$) and its derivatives forms on silica-alumina.[64] The fact that the absorption spectra characteristic of this radical do not appear when the usual mineral acids are mounted on silica seems to indicate the existence of electron-acceptor sites (L acid sites in the most general sense) on the silica-alumina surface. Hall,[65] Terenin *et al.*,[66] Roberts *et al.*,[67] Brouwer,[68] Rooney and Pink,[69] and Imelik *et al.*[70] found from electron spin resonance and absorption spectra that anthracene and perylene form cation radicals with silica-alumina, and attribute this oxidation process to L acid sites. Diphenylpicrylhydrazyl (DPPH) is also used as an electron donor reagent in the estimation of L acid amount.[71,72] However, we should maintain the distinction between those sites which accept one electron (L acids in the widest sense), and those sites which accept an electron pair (L acids properly so called).

Trambouze *et al.* measured the rise in temperature from the heat of reaction between a solid acid and dioxane or ethyl acetate in benzene.[73] It is doubtful, however, whether the assumption underlying their results (mentioned in 2.2.2) and those for similar work by Topchieva *et al.*[36] is justified, i.e. that ethyl acetate and dioxane react only with L acid.

Leftin and Hall have reported that the amount of L acid can be determined from the amount of triphenylcarbonium ions formed when the solid acid abstracts a hydride ion from triphenylmethane as below:

$$Ph_3CH + \underset{\text{L acid}}{\rceil\_\_\_\ulcorner} \longrightarrow Ph_3C^+ + \rceil\underline{H^-}\ulcorner$$

Carbonium ions, with absorption maximum at 420 mμ, are formed by boron trifluoride mounted on silica, but not by hydrogen fluoride on silica.[74] Again, the fact that no evolution of hydrogen takes place with B acids is a further indication that the carbonium ion cannot be formed by B acids, for if it were possible triphenylmethane would react according to the following formula,

$$Ph_3CH + H^+ \longrightarrow Ph_3C^+ + H_2\uparrow$$

with the consequent evolution of hydrogen. The method has been refined by Shiba *et al.*[75,76] However, this method has been subjected to criticism by Hirschler and Hudson on the basis of their experimental results with product analysis, photochemical effects, alkali poisoning etc.[77] These appear to indicate that oxidation sites can be identified in accordance with the formula

$$Ph_3CH + O(ad) \longrightarrow Ph_3COH$$
$$Ph_3COH + H^+ \longrightarrow Ph_3C^+ + H_2O$$

Porter and Hall recognized that carbonium ion formation cannot give a dependable measure of the amount of L acid because although carbonium ion formation is kinetically controlled, the stability of the carbonium ion from triphenylcarbinol is different from that of triphenylmethane.[78] Furthermore there is a photochemical oxidation of triphenylmethane. Recently Arai *et al.* have reported that the absorptions at 430 mμ and 340 mμ which appear when triphenylmethane is adsorbed on a solid are due to the carbonium ion formed by the reaction of triphenylmethane with B acid and the radical formed by the reaction with oxidizing sites respectively.[79]

It seems clear from the above-mentioned experiments that triphenylmethane is not only adsorbed on L acid sites, but also on both B acid and oxidizing sites. Therefore the amount of L acid cannot necessarily be obtained from the amount of the carbonium ion of triphenylmethane.

A method using xanthone which has been studied by Cook is useful for determining the acid strength of L acids.[80] Xanthone does not react with B acids, but reacts with L acids to cause a shift in the carbonyl infrared absorption spectra (the larger the shift, the stronger the acid). The order of the strengths of L acids obtained in this way is as follows; $BI_3 > BBr_3 > SbCl_5 > SbF_5 > BCl_3 > TiBr_4 > TiCl_4 > ZrCl_4 > PF_5 > AlBr_3 > AlCl_3 > FeBr_3 > FeCl_3 > BF_3 > SnBr_4 > SnCl_4 > InCl_3 > BiCl_3 > ZnCl_2 > HgBr_2 > HgCl_2 > CaBr_2 > CoCl_2 > CaCl_2 > CdCl_2$. The IR spectra of ethyl acetate are also used for the same purpose.[81] Saegusa *et al.* have applied the xanthone method to the measurement of L acidity of some metal complexes such as triethylaluminium-ethanol.[82]

There are other qualitative methods using indicators such as phenolphthalein, crystal violet and the leuco base of malachite green, which react with L acids to give different colours from those with B acids.[26,83] For example, phenolphthalein reacts with an L acid site on a solid surface by opening up its lactone ring and giving a distinctive red colour different from the purple which is usually observed in basic solutions.

2.3.3 Methods for the independent measurement of B and L acids

Infrared spectroscopic studies of ammonia and pyridine adsorbed on solid surfaces have made it possible to distinguish between B and L acid and to assess the amounts of B and L acid independently. The original work was reported in a paper by Mapes and Eischens, who found that the IR spectra of ammonia adsorbed on silica-alumina show two kinds of absorption, one of which indicates NH_3 adsorbed on L acid sites, and the other NH_4^+ on B acid sites.[84] Recently, Basila and Kantner have shown that the modes in which ammonia is adsorbed on silica-alumina are as physically adsorbed NH_3, as coordinately bonded NH_3, and as NH_4^+, each of which can be detected by means of their absorption bands (see Table 2–5).[85] Their investigations of the relative intensities of the corresponding bands showed a ratio of L to B acid sites of 4 : 1. The fact that the spectrum of pyridine coordinately bonded to the surface is very different from that of the pyridinium ion (see Table 2–6) also permits differentiation between acid types on the surface of a solid acid.[86] From the frequency shift of one of the bands of coordinately bonded pyridine as compared with that found in the liquid phase, and from the extent to which the band is retained upon evacuation and heating, a very rough estimate of surface L acid sites can be made. Parry showed that alumina has considerable strong L acidity but no B acidity, while silica-alumina has both.[86] Recently, piperidine, with a higher pK_a value than pyridine, has been used to

TABLE 2–5 Assignment of the bands of NH_3 chemisorbed on $SiO_2 \cdot Al_2O_3$

Frequency (cm^{-1})	Adsorbed species†	Assignment
3,341	LNH_3, PNH_3	ν_3 (e) (NH stretch)
3,280	LNH_3, PNH_3	ν_1 (a_1) (NH stretch)
3,230	NH_4^+	ν_3 (t_2) (NH stretch)
3,195	NH_4^+	ν_1 (a_1) (NH stretch)
1,620	LNH_3, PNH_3	ν_4 (l) (HNH deformation)
1,432	NH_4^+	ν_4 (t_2) (HNH deformation)

† LNH_3: NH_3 coordinately bonded to L acid site
PNH_3: physically adsorbed NH_3

TABLE 2–6 Infrared bands of pyridine on acid solids in the 1,400 ~ 1,700 cm^{-1} region†

Hydrogen bonded pyridine	Coordinately bonded pyridine	Pyridinium ion
1,440 ~ 1,447 (v.s.)	1,447 ~ 1,460 (v.s.)	
1,485 ~ 1,490 (w)	1,488 ~ 1,503 (v)	1,485 ~ 1,500 (v.s.)
		1,540 (s)
1,580 ~ 1,600 (s)	~ 1,580 (v)	
	1,600 ~ 1,633 (s)	~ 1,620 (s)
		~ 1,640 (s)

† Band intensities: v.s.–very strong; s–strong; w–weak; v–variable

differentiate B acid sites from L acid sites on zeolite catalysts.[87)]

Webb has found by spectroscopic studies on adsorbed diphenylethylene that carbonium ions (diphenylethyl cations) and charge transfer complexes are formed on a silica-alumina surface.[88)] The carbonium ions are considered to be formed by direct coordination with aluminium atoms (L acid sites) on silica-alumina, because the strong carbonium absorption band appears when the catalyst is dehydrated, and is similar to that observed with anhydrous aluminium chloride. The complex, on the other hand, is formed by the reaction with B acid sites, for the band associated with the complex appears even when the catalyst is hydrated.

There are many other methods aiding the correct classification of acid sites as B or L acid. Information about the active sites is yielded by both spectroscopic studies on reaction intermediaries, and the study of the exchange reaction of hydrogen between substrates and solid acid catalysts. Since these methods are rather specialized, only the references are cited here. The active sites for the reactions of olefin have been studied by Peri,[89)] Ozaki and Kimura[90)] and Webb,[88)] those for the reactions of paraffin by Thomas,[38)] Pines and Wackher,[91)] Lucchesi *et al.*,[92)] Greensfelder *et al.*[93)] and Hindin *et al.*[94)] and those for the reaction of alkyl aromatic compounds by Okuda and Tachibana,[64)] Peri[89)] and Johnson and Melik.[95)]

2.4 Relationship between acid strength and acid amount

The amount of acid at various acid strengths is obtained by the amine titration method described in 2.2.1, using the various basic indicators described in Table 2–1. The plots of the amount of acid against strength for silica-magnesia, silica-alumina, and Filtrol are shown in Fig. 2–12.[31)] The number of acid sites determined by using an indicator with $pK_a = +3.3$ is 0.85 mmol/g for silica-magnesia, which represents all the acid sites with $H_0 \leqq +3.3$. Similarly, the amount measured using an indicator of $pK_a = +1.5$ is 0.78 mmol/g, which represents all the acid sites with $H_0 \leqq +1.5$. Thus, the difference between these two acid amounts (0.85—0.78 mmol/g = 0.07 mmol/g) gives a measure of the acid sites in the range H_0 from +3.3 to +1.5. The acid amount in the same range of H_0 for silica-alumina is less than the above value for silica-magnesia, being only 0.02 mmol/g. Within the range of H_0 from +1.5 to −1 the difference is even more marked, for in silica-alumina acid amount drops to almost zero, but is very large (0.78 mmol/g) in silica-magnesia. On the other hand for values of $H_0 \leqq -8.2$

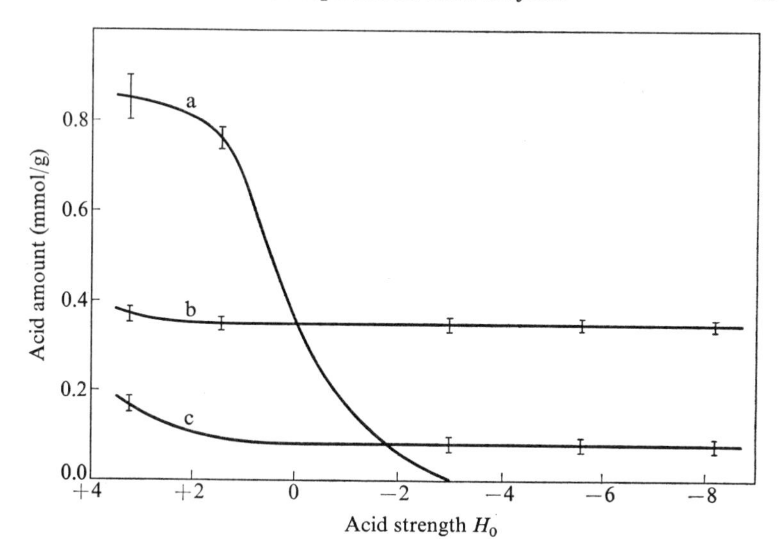

Fig. 2–12 Acid amount *vs.* acid strength for three solid acids
a: $SiO_2 \cdot MgO$ b: $SiO_2 \cdot Al_2O_3$ c: Filtrol

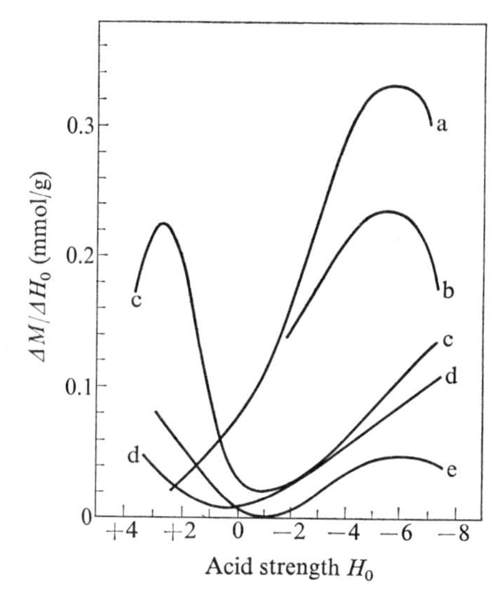

Fig. 2–13 Distribution of acid sites at different strengths
a: Al-III, $Al_2(SO_4)_3 \cdot xH_2O/SiO_2$ (wt ratio; 2/7),
b: Al-II, $Al_2(SO_4)_3 \cdot xH_2O/SiO_2$ (wt ratio; 1/7),
c: Zn-III, $ZnSO_4 \cdot xH_2O/SiO_2$ (wt ratio; 2/7),
d: Mn-III, $MnSO_4 \cdot xH_2O/SiO_2$ (wt ratio; 2/7),
e: silica gel

the acid amount in silica-magnesia is zero, but still fairly large in silica-alumina. This leads to the conclusion that silica-magnesia is catalytically inactive in reactions where a strong acid strength ($H_0 \leqq -3$) is required, but highly active for reactions requiring relatively weak acid strengths ($H_0 > -1$).

In Fig. 2–12 the acid amount at some arbitrary value of $H_0 =$ (say) a is actually that for the total acid amount with $H_0 \leqq a$. In order to display clearly the acid amount at $H_0 = a$, Ogino plotted $\Delta M/\Delta H_0$ (the difference in the amount of acid over a small range, ΔH_0, of H_0) against values of H_0 as shown in Fig. 2–13.[96] The figure clearly shows that at $H_0 \cong -5$ Al-III catalyst, curve a, has the largest acid amount, but that at $H_0 \cong +3$ it is largest in the Zn-III catalyst, curve c.

REFERENCES

1. C. Walling, *J. Am. Chem. Soc.*, **72,** 1164 (1950).
2. L. P. Hammett and A. J. Deyrup, *ibid.*, **54,** 2721 (1932); L. P. Hammett, *Chem. Rev.*, **16,** 67 (1935); *Physical Organic Chemistry*, chap. 9, McGraw-Hill, 1940.
3. A. E. Hirschler, *J. Catalysis*, **2,** 428 (1963).
4. H. Weil-Malherbe and J. Weiss, *J. Chem. Soc.*, **1948,** 2164.
5. K. Ikebe, N. Hara and K. Mita, *Kogyo Kagaku Zasshi*, **56,** 722 (1953).
6. H. A. Benesi, *J. Am. Chem. Soc.*, **78,** 5490 (1956).
7. K. Tanabe and R. Ohnishi, *J. Res. Inst. Catalysis, Hokkaido Univ.*, **10,** 229 (1962).
8. K. Tanabe and T. Yamaguchi, *ibid.*, **11,** 179 (1964).
9. K. Tanabe, C. Ishiya, I. Matsuzaki, I. Ichikawa and H. Hattori, *Dai 23-nenkai Kōenyokō-shū* (Japanese) (Ann. Meeting Chem. Soc. Japan, 23rd, Tokyo, Preprints of Papers), No. 03408 (1970).
10. H. P. Leftin and M. C. Hobson, Jr., *Advances in Catalysis*, vol. 14, p. 115, Academic Press, 1963.
11. A. N. Terenin, *ibid.*, vol. 15, p. 227, 1964.
12. N. S. Kotsarenko, L. G. Karakchiev and V. A. Dzisko, *Kinetika i Kataliz*, **9,** 158 (1968).
13. J. Kobayashi, *Nippon Kagaku Zasshi*, **84,** 21, 25 (1963); **82,** 288 (1961); **80,** 1399 (1959).
14. J. Kobayashi and I. Higuchi, *Shokubai* (*Tokyo*), **10,** (23rd Symp. Catalysis, Preprints of Papers), 123 (1968).
15. A. N. Webb, *Ind. Eng. Chem.*, **49,** 261 (1957).
16. E. E. Roper, *Discussions Faraday Soc.*, **8,** 270 (1950).
17. H. Bremer and K. H. Steinberg, *Intern. Congr. Catalysis, 4th, Moscow, Preprints of Papers,* No. 76 (1968).
18. R. L. Richardson and S. W. Benson, *J. Phys. Chem.*, **61,** 405 (1957).
19. A. C. Zettlemoyer and J. J. Chessick, *ibid.*, **64,** 1131 (1960).
20. R. L. Stone and H. F. Rase, *Anal. Chem.*, **29,** 1273 (1957).
21. P. L. Hsieh, *J. Catalysis*, **2,** 211 (1963).

22. V. Kevorkian and R. O. Steiner, *J. Phys. Chem.*, **67**, 545 (1963).
23. A. Clark, V. C. F. Holm and D. M. Blackburn, *J. Catalysis*, **1**, 244 (1962); A. Clark and V. C. F. Holm, *ibid.*, **2**, 16, 21 (1963).
24. A. J. de Rosset, C. G. Finstron and C. J. Adams, *ibid.*, **1**, 235 (1962).
25. M. Misono, Y. Saito and Y. Yoneda, *Proc. Intern. Congr. Catalysis, 3rd, Amsterdam*, I, No. 18 (1964).
26. H. Pines and W. O. Haag, *J. Am. Chem. Soc.*, **82**, 2471 (1960).
27. T. Aonuma, M. Sato and T. Shiba, *Shokubai* (*Tokyo*), **5**, No. 3, 274 (1963).
28. M. W. Tamele, *Discussions Faraday Soc.*, **8**, 270 (1950).
29. K. Tanabe and M. Katayama, *J. Res. Inst. Catalysis, Hokkaido Univ.*, **7**, 106 (1959).
30. O. Johnson, *J. Phys. Chem.*, **59**, 827 (1955).
31. H. A. Benesi, *ibid.*, **61**, 970 (1957).
32. I. Matsuzaki, Y. Fukuda, T. Kobayashi, K. Kubo and K. Tanabe, *Shokubai* (*Tokyo*), **11**, No. 6, 210 (1969).
33. K. Tanabe and Y. Watanabe, *J. Res. Inst. Catalysis, Hokkaido Univ.*, **11**, 65 (1963).
34. S. E. Voltz, A. E. Hirschler and A. Smith, *J. Phys. Chem.*, **64**, 1594 (1960).
35. Y. Trambouze, *Compt. Rend.*, **233**, 648 (1951); Y. Trambouze, L. de Mourgues and M. Perrin, *J. Chim. Phys.*, **51**, 723 (1954); *Compt. Rend.*, **236**, 1023 (1953).
36. K. V. Topchieva, I. F. Moskovskaya and N. A. Dobrokhotova, *Kinetics and Catalysis* (*USSR*) (*Eng. Transl.*), **5**, 910 (1964).
37. K. Tanabe and T. Yamaguchi, *J. Res. Inst. Catalysis, Hokkaido Univ.*, **14**, 93 (1966).
38. C. L. Thomas, *Ind. Eng. Chem.*, **41**, 2564 (1949).
39. G. A. Mills, E. R. Boedecker and A. G. Oblad, *J. Am. Chem. Soc.*, **72**, 1554 (1950).
40. T. H. Milliken, Jr., G. H. Mills and A. G. Oblad, *Discussions Faraday Soc.*, **8**, 279 (1950).
41. E. Echigoya, *Nippon Kagaku Zasshi*, **76**, 1049 (1955).
42. H. Uchida and M. Temma, *Shokubai* (*Tokyo*), **4**, No. 4, 353 (1962).
43. Y. Tezuka and T. Takeuchi, *Bull. Chem. Soc. Japan*, **38**, 485 (1965).
44. Y. Murakami and T. Shiba, *Actes Congr. Intern. Catalyse*, 2[e], *Paris*, III, No. 129 (1960, Pub. 1961).
45. Y. Kubokawa, *J. Phys. Chem.*, **67**, 769 (1963); Y. Kubokawa and S. Toyama, *Shokubai* (*Tokyo*), **4**, No. 1, 46 (1962).
46. T. Shirasaki, M. Mimura and K. Mukaida, *Bunseki Kiki* (Japanese), **5**, No. 7, 59 (1968).
47. J. E. Benson, K. Ushiba and M. Boudart, *J. Catalysis*, **9**, 91 (1967).
48. F. S. Stone and L. Whalley, *ibid.*, **8**, 173 (1967).
49. R. T. Barth and E. V. Ballou, *Anal. Chem.*, **33**, 1080 (1961).
50. Y. Amenomiya, J. H. B. Chenier and R. J. Cvetanović, *J. Phys. Chem.*, **68**, 52 (1964).
51. Y. Murakami, H. Nozaki and J. Turkevich, *Shokubai* (*Tokyo*), **5**, No. 3, 262 (1963).
52. A. G. Oblad, T. H. Milliken, Jr. and G. A. Mills, *Advances in Catalysis*, vol. 3, p. 202, Academic Press, 1951.
53. K. Tarama, S. Teranishi, K. Hattori and T. Ishibashi, *Shokubai* (*Tokyo*), **4**, No. 1, 69 (1962).
54. P. Stright and J. D. Danforth, *J. Phys. Chem.*, **57**, 448 (1953).

55. N. Ohta, *Kogyo Kagaku Zasshi*, **51,** 16 (1948).
56. E. Jungreis and L. Ben-Dor, *Israel J. Chem.*, **3,** 1 (1966).
57. A. Grenhall, *Ind. Eng. Chem.*, **41,** 1485 (1949).
58. C. J. Plank, *Anal. Chem.*, **24,** 1304 (1952).
59. K. A. Mähl, *Ind. Eng. Chem., Anal. Ed.*, **12,** 24 (1940).
60. Y. Trambouze, L. de Mourgues and M. Perrin, *Compt. Rend.*, **234,** 1770 (1952).
61. V. C. F. Holm, G. C. Bailey and A. Clark, *J. Phys. Chem.*, **63,** 129 (1959).
62. J. D. Danforth, *Actes Congr. Intern. Catalyse*, 2^e, *Paris*, II, No. 61 (1960, Pub. 1961).
63. S. Malinowski and S. Szczepanska, *J. Catalysis*, **2,** 310 (1963).
64. N. Okuda and T. Tachibana, *Bull. Chem. Soc. Japan*, **33,** 863 (1960); *Shokubai (Tokyo)*, **4,** No. 1, 54 (1962); N. Okuda, *Nippon Kagaku Zasshi*, **82,** 1290 (1961).
65. W. K. Hall, *J. Catalysis*, **1,** 53 (1962).
66. A. N. Terenin, E. Kotov, V. Barachevsky and V. Holmogorov, *Intern. Symp. Mol. Structure Spectry., Science Council of Japan, Tokyo, Sept.* 10 ~ 15, *Preprints of Papers*, D105 (1962).
67. R. M. Roberts, C. Barter and H. Stone, *J. Phys. Chem.*, **63,** 2077 (1959).
68. D. M. Brouwer, *Chem. Ind. (London)*, 177 (1961).
69. J. J. Rooney and R. C. Pink, *Proc. Chem. Soc.*, 70 (1961).
70. Y. Kodratoff, C. Naccache and B. Imelik, *J. Chim. Phys.*, **65,** 562 (1968).
71. T. Kawaguchi, S. Hasegawa and Y. Nakamura, *Tokyo Gakugei Daigaku Kiyō* (Japanese), **20,** Section 4, No. 1, p. 75 (1968).
72. T. Kawaguchi, S. Hasegawa, K. Kaseda and S. Kurita, *Denshi Shashin* (Japanese), **8,** 92 (1968).
73. Y. Trambouze, M. Perrin and L. de Mourgues, *Advances in Catalysis*, vol. 9, p. 544, Academic Press, 1957.
74. H. P. Leftin and W. K. Hall, *Actes Congr. Intern. Catalyse, 2^e, Paris*, II, No. 65 (1960, Pub. 1961).
75. A. Nakamura, K. Sano, K. Takemura and T. Shiba, *Shokubai (Tokyo)*, **4,** No. 1, 58 (1962).
76. M. Sato, H. Hattori, K. Yoshida and T. Shiba, *ibid.*, **6,** No. 2, 80 (1964).
77. A. E. Hirschler and J. O. Hudson, *J. Catalysis*, **3,** 239 (1964).
78. R. P. Porter and W. K. Hall, *ibid.*, **5,** 366 (1966).
79. H. Arai, Y. Saito and Y. Yoneda, *Bull. Chem. Soc. Japan*, **40,** 312 (1967).
80. D. Cook, *Can. J. Chem.*, **41,** 522 (1963).
81. M. F. Lappert, *J. Chem. Soc.*, **1962,** 542.
82. T. Saegusa, H. Imai, T. Uejima and J. Furukawa, *Shokubai (Tokyo)*, **7,** No. 1, 43 (1965).
83. G. Kortüm, J. Vogel and W. Braun, *Angew. Chem.*, **70,** 651 (1958).
84. J. E. Mapes and R. R. Eischens, *J. Phys. Chem.*, **58,** 809 (1954); W. A. Pliskin and R. R. Eischens, *ibid.*, **59,** 1156 (1955).
85. M. R. Basila and T. R. Kantner, *ibid.*, **71,** 467 (1967).
86. E. P. Parry, *J. Catalysis*, **2,** 371 (1963).
87. T. R. Hughes and H. M. White, *J. Phys. Chem.*, **71,** 2192 (1967).
88. A. N. Webb, *Actes Congr. Intern. Catalyse, 2^e, Paris*, II, No. 62 (1960, Pub. 1961).
89. J. B. Peri, *ibid.*, II, No. 64 (1960, Pub. 1961).
90. A. Ozaki and K. Kimura, *J. Catalysis*, **3,** 395 (1964); *Shokubai (Tokyo)*, **4,** No. 1, 61 (1962).
91. H. Pines and R. C. Wackher, *J. Am. Chem. Soc.*, **68,** 595 (1946).

92. P. J. Lucchesi, D. L. Breader and J. P. Longwell, *ibid.*, **81,** 3235 (1959).
93. B. S. Greensfelder, H. H. Voge and G. M. Good, *Ind. Eng. Chem.*, **39,** 1032 (1947).
94. S. G. Hindin, G. A. Mills and A. G. Oblad, *J. Am. Chem. Soc.*, **73,** 278 (1951).
95. M. F. L. Johnson and J. S. Melik, *Ann. Meeting Am. Chem. Soc., Div. Petrol. Chem., Preprints of Papers*, vol. 5, B–109 (1960).
96. Y. Ogino, *Shokubai* (*Tokyo*), **4,** No. 1, 73 (1962).

Chapter 3

Determination of Basic Properties on Solid Surfaces

Little work has been done on the measurement of basic properties, but the methods are similar to those described in the foregoing chapter.

3.1 Basic strength

The basic strength of a solid surface is defined as the ability of the surface to convert an adsorbed electrically neutral acid to its conjugate base, i.e. the ability of the surface to donate an electron pair to an adsorbed acid. There are two methods for the measurement of basic strength; the method using indicators, and the phenol vapour adsorption method.

3.1.1 Indicator method

When an electrically neutral acid indicator is adsorbed on a solid base from a non-polar solution, the colour of the acid indicator is changed to that of its conjugate base, provided that the solid has the necessary basic strength to impart electron pairs to the acid. Thus it is generally possible to determine the basic strength by observing the colour changes of acid indicators over a range of pK_a values.

For the reaction of an indicator AH with a solid base $\overline{\mathrm{B}}$,

$$\mathrm{AH} + \overline{\mathrm{B}} \rightleftharpoons \mathrm{A}^- + \overline{\mathrm{B}}\mathrm{H}^+$$

the basic strength H_0 of $\overline{\mathrm{B}}$ is given by an equation similar to Eq. 2.2,

$$H_0 = \mathrm{p}K_a + \log[\mathrm{A}^-]/[\mathrm{AH}] \tag{3.1}$$

where [AH] is the concentration of the acidic form of the indicator, and $[\mathrm{A}^-]$ the concentration of the basic form.

The first perceptible change in the colour of an acid indicator occurs when about 10% of the adsorbed layer of indicator is in the basic form, i.e. when the ratio $[\mathrm{A}^-]/[\mathrm{AH}]$ reaches $0.1/0.9 \cong 0.1$.

Further increase in the intensity of the colour is only perceptible to the naked eye when about 90% of the indicator is in the basic form, i.e. $[A^-]/[AH] = 0.9/0.1 \cong 10$. Thus the initial colour change and the subsequent change in intensity are observed at values of H_0 equal to pK_a-1 and pK_a+1 respectively. If we assume that the intermediate colour appears when the basic form reaches 50%, i.e. when $[A^-]/[AH]=1$, we have $H_0=pK_a$.

According to this assumption, the approximate value of the basic strength on the surface is given by the pK_a value of the adsorbed indicator at which the intermediate colour appears.[1] Indicators which lend themselves to this method are listed in Table 3–1. Non-polar solvents such as benzene and isooctane are used for the indicators.

TABLE 3–1 Indicators used for the measurement of basic properties[2,3]

Indicators	Colour		pK_a
	Acid-form	Base-form	
Bromothymol blue	yellow	green	7.2
2,4,6-Trinitroaniline	yellow	reddish-orange	12.2
2,4-Dinitroaniline	yellow	violet	15.0
4-Chloro-2-nitroaniline	yellow	orange	17.2
4-Nitroaniline	yellow	orange	18.4
4-Chloroaniline	colourless	pink†1	26.5†2

†1 The colour disappears with the addition of benzoic acid
†2 This value was estimated from the data of Stewart and Dolman[4]

This method gave $H_0 \leqq +4$ as the basic strength of a catalyst with 20 mmol Na to 100 g of silica gel when it was dried at 120 °C, but a value of $H_0 \leqq +9.3$ when it had been heated to 1,000 °C. It has also been shown that a higher sodium content (80 mmol Na per 100 g silica gel) will change phenolphthalein to its red form even when dried at only 120 °C.[1] Other solids on which adsorbed bromothymol blue shows its characteristic colour are CaO, MgO, Al_2O_3, Na_2CO_3, K_2CO_3, $KHCO_3$, $(NH_4)_2CO_3$, $BaCO_3$, $SrCO_3$, $5ZnO \cdot 2CO_3 \cdot 4H_2O$, $KNaCO_3$, $Na_2WO_4 \cdot 2H_2O$ and KCN.[5,6]

Malinowski *et al.*, by using bromocresol purple ($pK_a=+6.0$), methyl red (+4.8), and bromophenol blue (+3.8) in addition to the acid indicators mentioned above, were able to measure the acid-base strength of magnesium oxide prepared by heating the hydroxide.[7] The results are shown in Fig. 3–1; basic strength decreases and acid

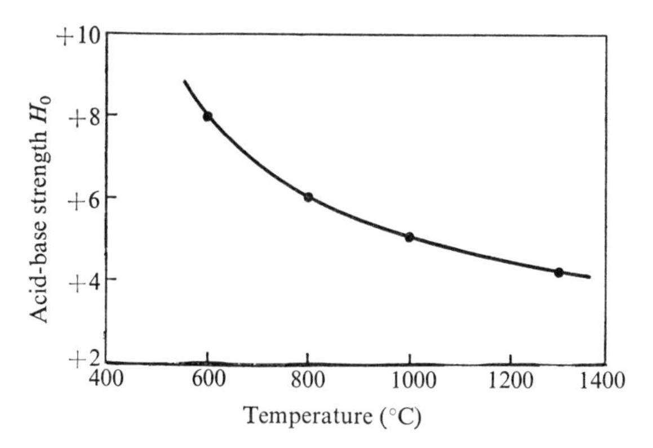

Fig. 3–1 Acid-base strength of magnesium oxide *vs.* heat-treatment temperature[7]

strength increases as the temperature of heat-treatment is raised.

There are, in addition, a series of UV spectroscopic studies which have been made for 1,3,5-trinitrobenzene adsorbed on magnesium oxide,[8] *o*-nitrophenol adsorbed on alkaline earth metal oxides,[9,10] three mononitrophenols adsorbed on alkali metal carbonates,[11] and *p*-nitrophenol, phenolphthalein and thymolphthalein adsorbed on porous calcium fluoride and barium fluoride film.[12–14] In these studies the absorption maxima of the transmission or reflection spectra may be correlated with the basic strength of the solid in its widest sense, that is the strength of single electron donor sites. The amount of these single electron donor sites on alumina has been estimated from the signal intensity of the ESR spectra of anion radicals formed in the reaction of tetracyanoethylene with alumina.[15]

3.1.2 Phenol vapour adsorption method

Phenol, a weak acid which is stable at relatively high temperatures, is used as an adsorbate. The phenol vapour is adsorbed on the solid at a given vapour pressure at a constant temperature, the system being evacuated at elevated temperatures (100, 200, 300 and 380 °C) after saturation point has been reached. At each temperature the amount of adsorbate retained on the solid surface is measured. A solid on which the adsorbed phenol is difficult to desorb even at the higher temperatures is said to have high basic strength. The classification of certain metal oxides according to their basic strengths is as follows: 1) strong; CaO, 2) moderate; BeO, MgO, ZnO, 3) weak; silica gel, alumino-silicate.[16]

3.2 Amount of base

The "amount" of base (basic sites) on a solid is usually expressed as the number (or mmol) of basic sites per unit weight or per unit surface area of the solid. It is also sometimes more loosely called "basicity." Methods for determining the amount of base may be divided into the following four types.

3.2.1 Titration method

Just as the acid amount is determined by *n*-butylamine titration (2.2.1), the amount of basic sites can be measured by titrating a suspension in benzene of a solid on which an indicator has been adsorbed in its conjugate basic form, with benzoic acid dissolved in benzene. The benzoic acid titres are a measure of the amount of basic sites (in mmol/g or mmol/m^2) having a basic strength corresponding to the pK_a value of the indicator used.[3]

Fig. 3–2 shows the variation in base amount for a typical metal oxide, CaO, with bromothymol blue indicator ($pK_a = +7.1$), as determined by this method.[17] The base amount of alkaline earth metal oxides is changed by heat-treatment, that of $Ca(OH)_2$ showing a particularly sharp increase with increasing temperature of heat-treatment,

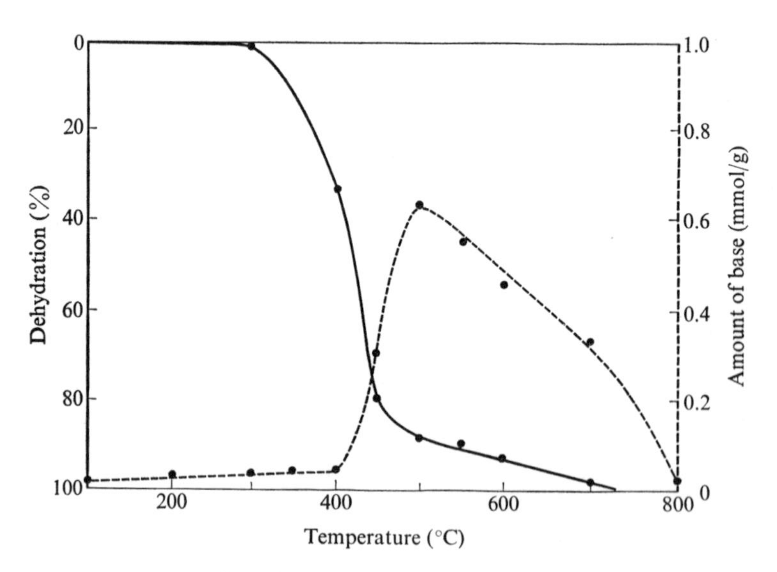

Fig. 3–2 Effect of heat-treatment temperature on $Ca(OH)_2$
(--●--) amount of base, (—●—) percentage dehydration

reaching a maximum at 500 °C before beginning to decrease. By applying the same method to slags such as $SiO_2 \cdot CaO \cdot MgO$, $SiO_2 \cdot CaO \cdot TiO_2$, and $CaO \cdot Al_2O_3 \cdot MgO$, Niwa *et al.* were able to measure the amounts of base, with the results shown in Table 3–2.[18] The basicity of the slags numbered 1 to 9 in the Table is in the order 1, 2, 4, 6, 5, 3, 7, 9, and 8, while the values calculated from Mori's expression for slag basicity[19] are in the order 1, 4, 2, 6, 3, 5, 7, 9, and 8. There is thus very close agreement between the theoretical and experimental orders of basicity, the only exceptions being the reversals between 2 and 4, and 3 and 7.

TABLE 3–2 Basicity of $SiO_2 \cdot CaO \cdot MgO$

No.	Composition of slag (mol%)			Basicity (mmol/g × 10^3)	BL†
	SiO_2	CaO	MgO		
1	44.2	55.7	0	1.37	5.581
2	44.8	32.3	22.9	1.38	0.043
3	45.2	27.1	27.7	1.14	−0.105
4	46.0	39.3	14.7	1.27	0.063
5	46.8	33.7	19.5	1.22	−0.134
6	46.5	34.1	19.4	1.25	−0.095
7	51.3	36.3	12.4	0.44	−0.555
8	51.0	15.0	33.9	−6.2	−0.855
9	52.7	32.4	14.9	−5.7	−0.782

† BL (basicity) = $\sum a_i N_i$, where N_i is the mole fraction of each component and a_i the constant proper to each component[19]

This benzoic acid titration method, again with bromothymol blue as the indicator, reveals a very interesting correlation between the amount of water adsorbed on alumina and the amount of base. As shown in Fig. 3–3, the base amount increases sharply with increase in the amount of water within the range of 3 to 8 mmol/g and then attains a constant value.[20] The basic sites begin to appear when the adsorbed water is sufficient to poison all the acid sites on the alumina. No such appearance of basic sites due to water adsorption has ever been reported with either silica gel or silica-alumina.

Take *et al.* extended the range of indicators available for the measurement of solid basicity as shown in Table 3–1, and determined both the strength and the amount of base for several alkaline earth metal oxides[2] (see also 4.1).

Malinowski and Szczepanska have devised several titration methods for use with aqueous solutions and with anhydrous acetic acid

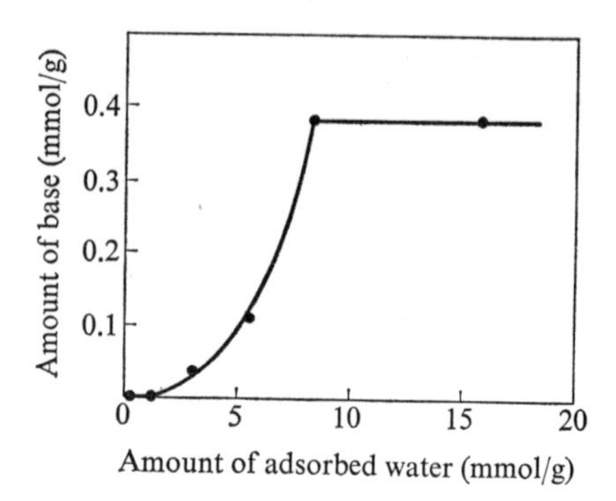

Fig. 3–3 Amount of base *vs*. amount of water adsorbed on Al_2O_3 at room temperature

solution.[1] The former consist of titration of a catalyst suspension in water against sulfuric acid solution, and the latter involves potentiometric titration with perchloric acid solution in anhydrous acetic acid. Fig. 3–4 shows the relationship which the former method has revealed between the amount of base on silica impregnated with sodium, and the amount of sodium adsorbed on the silica. In this case we have Arrhenius basicity due to the presence of OH^- ions formed in the solution through the hydrolysis of the Si–ONa compound (the salt of a weak acid and a strong base) to Si–OH and NaOH. However, as has been pointed out in 2.2.3, those methods which involve aqueous solutions require further investigation before they can be used with any great degree of confidence.

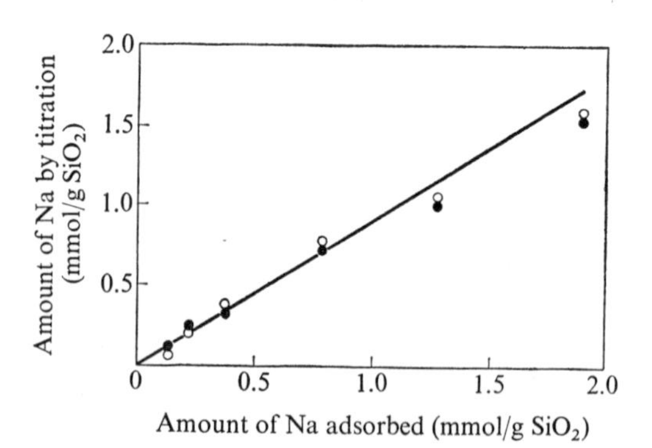

Fig. 3–4 Amount of Na found by titration in water (Arrhenius basicity) *vs*. amount of Na deposited on silica

(○) catalyst dried at 120 °C, (●) heat treated at 1,000 °C

3.2.2 Exchange method

Just as the acid amount may be determined by measuring the extent to which protons are released in solution by exchanges at the surface between protons and cations such as NH_4^+, K^+ etc. (see 2.3.1), so the base amount of some solids can be estimated from the exchanges taking place between hydroxide ions and anions at the surface. Naruko has measured the base amount of carbon activated with nitrous oxide[21] or ammonia[22] by means of a pH meter. He observed the change in pH of 100 ml of aqueous 0.1 N KCl solution (initially pH=7), to which 0.1 g of active carbon had been added, after 24 h of continuous shaking. The basicity of charcoal activated with nitrous oxide is listed in Table 3–3. It increases with increasing temperature of heat-treatment,

TABLE 3–3 Basicity, surface area and catalytic activity of nitrous oxide-activated charcoal

Temperature of heat-treatment (°C)	Surface area (m^2/g)	Basicity pH	Rate of decomposition of hydrogen peroxide (ml/day)
untreated	264	7.1	2
300	164	7.1	75
400	187	7.2	33
500	210	7.9	53
600	219	9.0	93
700	453	9.8	372
800	636	9.9	721
900	860	8.8	713

reaches a maximum value at 800 °C, and then decreases at higher temperatures.

3.2.3 Acid adsorption method

The amount of basic sites on pure sugar charcoal has been obtained by measuring the adsorbed amounts of such acids as benzoic, acetic and hydrochloric acids in water.[23] As Fig. 3–5 shows, the basicity attains a maximum value when the charcoal has been heat-treated at about 900 °C, while the acidity maximum associated with adsorption of NH_4OH and NaOH appears at 400 °C.

Recently Schwab has found that boron trifluoride (a Lewis acid) covers from 60 to 80% BET surface area of alumina at 30 mmHg pressure and 400 °C.[24] Since ammonia covers only from 5 to 8% for the

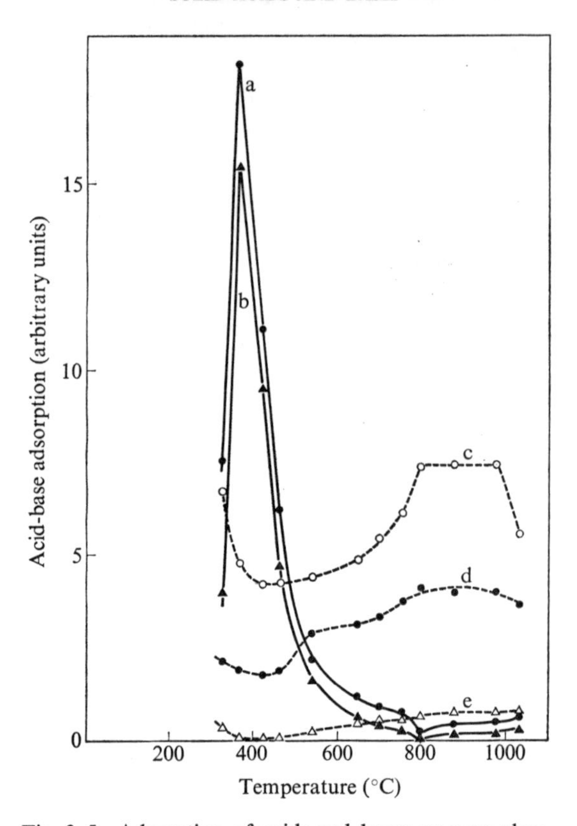

Fig. 3–5 Adsorption of acids and bases on pure charcoal *vs.* activation temperature
a (—●—) NH_4OH, b (—▲—) NaOH, c (--○--) C_6H_5COOH, d (--●--) CH_3COOH, e (--△--) HCl

same sample of alumina, the basicity of alumina is greater than its acidity.

The number of carbon dioxide molecules adsorbed per unit surface area may also be considered a measure of the amount of basic sites on the surface.[25]

3.2.4 Calorimetric titration

The calorimetric determination of acid amount on a surface was described in 2.2.2. Similarly the base amount can be determined by observing the rise in temperature due to the heat of reaction between solid bases and acid in benzene. A titration curve for the basicity measurement of silica-alumina is given in Fig. 3–6.[26] Taking 3 ml

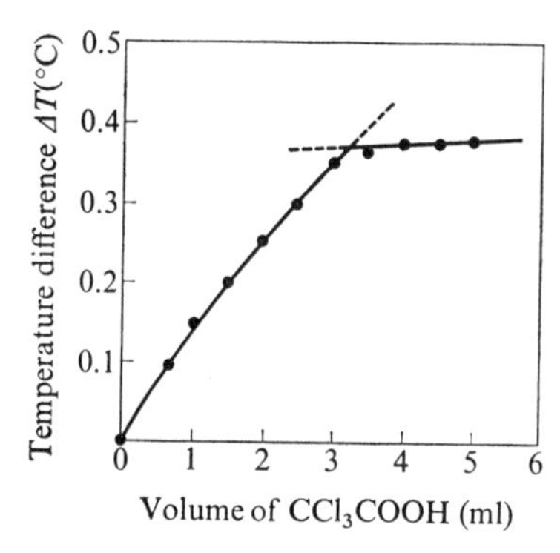

Fig. 3–6 Calorimetric titration curve: 4.15 g $SiO_2 \cdot Al_2O_3$ with trichloroacetic acid in benzene

as the amount of trichloroacetic acid shown by the curve to be necessary for the neutralization of the catalyst's basic sites, the basicity is estimated to be about 0.6 mmol/g. However, it should be noted that an attempt to measure the basicity of silica-alumina by titration with benzoic acid using bromothymol blue indicator failed because the adsorbed indicator did not assume its basic colour.[3)]

REFERENCES

1. S. Malinowski and S. Szczepanska, *J. Catalysis*, **2**, 310 (1963).
2. J. Take, N. Kikuchi and Y. Yoneda, *Shokubai* (*Tokyo*), **10** (23rd Symp. Catalysis, Preprints of Papers), 127 (1968).
3. K. Tanabe and T. Yamaguchi, *J. Res. Inst. Catalysis, Hokkaido Univ.*, **11,** 179 (1964).
4. R. Stewart and D. Dolman, *Can. J. Chem.*, **45,** 925 (1967).
5. K. Tanabe and M. Katayama, *J. Res. Inst. Catalysis, Hokkaido Univ.*, **7,** 106 (1959).
6. K. Nishimura, *Nippon Kagaku Zasshi*, **81,** 1680 (1960).
7. S. Malinowski, S. Szczepanska, A. Bielanski and J. Sloczynski, *J. Catalysis*, **4,** 324 (1965).
8. G. Kortüm, *Angew. Chem.*, **70,** 651 (1958).
9. H. Zeitlin, R. Frei and M. McCarter, *J. Catalysis*, **4,** 77 (1965).
10. H. E. Zaugg and A. D. Schaffer, *J. Am. Chem. Soc.*, **87,** 1857 (1965).
11. H. Zeitlin, N. Kondo and W. Jordan, *J. Phys. Chem. Solids*, **25,** 641 (1964).
12. J. H. de Boer, *Z. Physik. Chem.*, **B16,** 397 (1932); **B17,** 161 (1932); J. H. de Boer and J. F. H. Custers, *ibid.*, **B25,** 225, 238 (1934); J. F. H. Custers and J. H. de Boer, *Physica*, **1,** 265 (1934); **3,** 407 (1936).
13. J. H. de Boer and G. M. M. Houben, *Koninkl. Ned. Akad. Wetenschap., Proc., Ser.* **B54,** 421 (1951).
14. A. N. Terenin, *Advances in Catalysis*, vol. 15, p. 227, Academic Press, 1964.

15. C. Naccache, Y. Kodratoff, R. C. Pink and B. Imelik, *J. Chim. Phys.*, **63,** 341 (1966).
16. O. V. Krylov and E. A. Fokina, *Probl. Kinetiki i Kataliza, Acad. Nauk, USSR*, vol. 8, p. 248, 1955.
17. K. Saito and K. Tanabe, *Shokubai* (*Tokyo*), **11,** No. 4, 206P (1969).
18. K. Niwa, S. Kado and H. Kuki, *Report of 19th Committee of Japan Soc. Promotion Sci.*, No. 6673 (1962).
19. K. Mori, *Tetsu To Hagane*, **46,** 466 (1960); *Nippon Kinzoku Gakkaishi*, **24,** 383 (1961).
20. M. Yamadaya, K. Shimomura, T. Konoshita and H. Uchida, *Shokubai* (*Tokyo*), **7,** No. 3, 313 (1965).
21. E. Naruko, *Kogyo Kagaku Zasshi*, **67,** 2019 (1964).
22. E. Naruko, *ibid.*, **67,** 2023 (1964).
23. A. King, *J. Chem. Soc.*, **1937,** 1489.
24. G.-M. Schwab and H. Kral, *Proc. Intern. Congr. Catalysis, 3rd, Amsterdam*, I, No. 20 (1964).
25. S. Malinowski, S. Szczepanska and J. Sloczynski, *J. Catalysis*, **7,** 67 (1964).
26. K. Tanabe and T. Yamaguchi, *J. Res. Inst. Catalysis, Hokkaido Univ.*, **14,** 93 (1966).

Chapter 4

Acid and Base Centres: Their Structure and Acid-Base Properties

The amount, strength, and kind of acid and base sites on various solid surfaces which have been determined by the methods described in Chapters 2 and 3 are given here in detail, and discussed in connection with the respective site structures.

4.1 Metal oxides and sulfides

The metal oxides and sulfides whose acidic and/or basic properties have been determined are Al_2O_3, SiO_2, MgO, CaO, SrO, BaO, ZnO, TiO_2, V_2O_5, Sb_2O_5, Cr_2O_3, MoO_3, As_2O_3, CeO_2, ZnS and CaS. Structural studies of acid and base centres have been made only for alumina, magnesia and calcium oxide.

4.1.1 Alumina

In general, alumina shows strong surface acidity when heated to temperatures above 470 °C in a vacuum.[1] The acid amounts and strengths of a commercial activated alumina sample heat-treated in air at 500 °C are given in Table 4–1.[2] The effect of calcination temperature on the acidic properties of pure alumina prepared from aluminium isopropoxide is illustrated in Fig. 4–1, which gives the acid amounts at various acid strengths H_0 as measured by the amine titration method (2.2.1), plotted against calcination temperature.[3] Two acidity maxima are to be seen at 500 °C and at 700∼800 °C, with an unusual acidity minimum at 600 °C. A wide variety of other techniques have also been used in the study of the acidity of alumina. These include titration with aqueous potassium hydroxide[4] (2.2.3), calorimetric titration with dioxane[5] (2.2.2) and the chemisorption of gaseous ammonia,[4,6] trimethylamine,[7] and pyridine[8] (2.2.4). Each of these methods gives acidity values for alumina which apparently approximate those of silica-alumina. Hirschler has measured the amount of NH_3 chemisorbed on a commercial Baker-Sinclair platinum-alumina reforming catalyst as a function of the temperature of evacuation.[6] The catalyst, activated

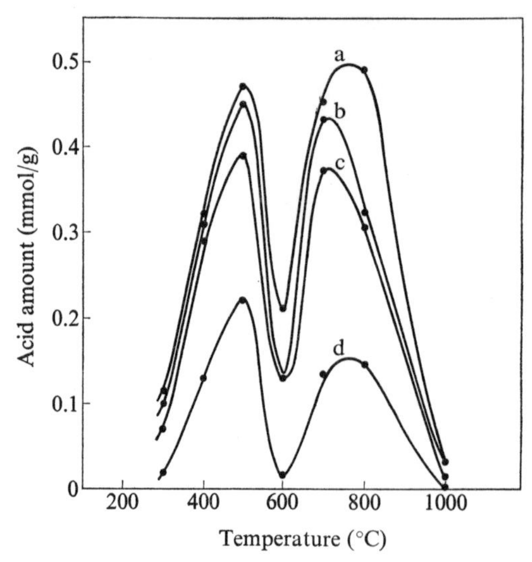

Fig. 4–1 Amount of acid on Al_2O_3 *vs.* calcination temperature at various acid strengths
a: acid strength $H_0 \leqq +3.3$ b: $H_0 \leqq +1.5$
c: $H_0 \leqq -3.0$ d: $H_0 \leqq -5.6$
X-ray analysis: 450∼500°C, η–Al_2O_3 (low crystallinity); 600 °C, η–Al_2O_3 (high crystallinity); 800 °C, η–Al_2O_3 + θ-Al_2O_3; 1,000 °C, α-Al_2O_3

by evacuation at 450 °C, was saturated with NH_3 at 0 °C, and then pumped continuously overnight to a pressure of 10^{-5} mmHg, successively raising the temperature by various discrete amounts. The amount of chemically adsorbed NH_3 was measured gravimetrically, with the results shown in Table 4–2. The negative "adsorption" at 450 °C indicates that the presence of the NH_3 catalyzed an additional loss of water of constitution. Pines and Haag have used the method detailed in

TABLE 4–1 Acidity of alumina†

Acid strength	$H_0 \leqq -5.6$	$H_0 \leqq -3.0$	$H_0 \leqq +1.5$	$H_0 \leqq +3.3$
Acid amount (mmol/g)	0	0.223	0.287	0.309

† Commercial activated alumina heat-treated in air at 500°C for 3 h

TABLE 4–2 Ammonia chemisorption by alumina as a function of temperature

Evacuation temperature (°C)	30	100	200	300	450
NH_3 adsorption (mmol/g)	1.09	0.66	0.40	0.18	−0.04

2.1.4 to estimate both the acid strength and the acid amount of an alumina sample.[7] They determined the upper limits of the total number of acid sites capable of dehydrating 1-butanol, and of the number of strong acid sites capable of isomerizing cyclohexane, as 10×10^{12} and 8×10^{12} sites per cm^2 respectively.

The problem of whether the acid sites of alumina are Brønsted or Lewis type has been intensively investigated by the various methods described in 2.3. Pines *et al.*, using a range of indicators (see 2.3.2), concluded that alumina displayed Lewis acidity.[7] Experiments based on the infrared absorption bands of pyridine adsorbed on alumina provide additional very strong evidence that alumina acid sites are of the Lewis type. The absorption spectra of pyridine adsorbed on dehydrated η-Al_2O_3, observed by Parry, are shown in Fig. 4–2.[9] The bands are

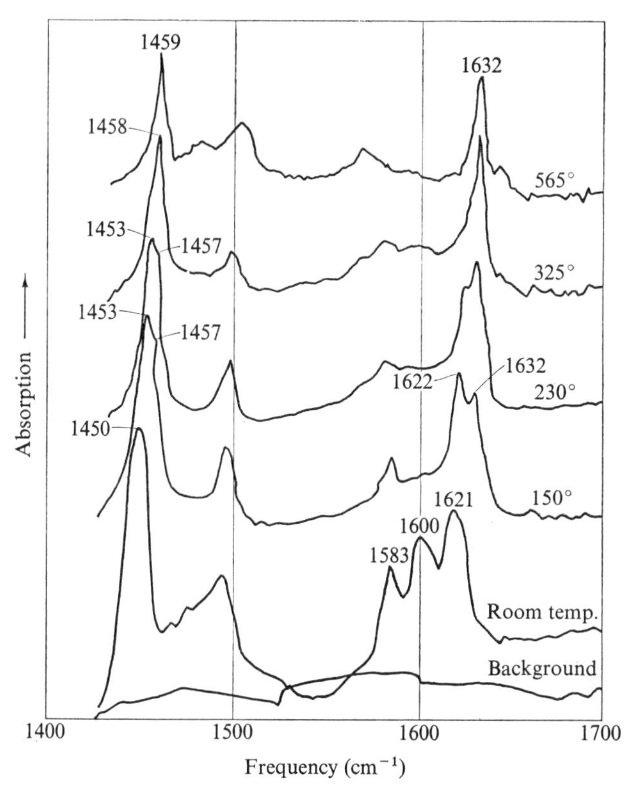

Fig. 4–2 Infrared absorption spectra for pyridine on η–Al_2O_3 The background level was obtained after evacuation of the sample at 450 °C for 3 h. Subsequent to pyridine addition, each spectrum was recorded after evacuation for 3 h at the temperature indicated (°C)

assigned as in Table 2–6. The retention of pyridine after evacuation at high temperatures, as evidenced by the persistence and frequency shift of the 1,459 cm^{-1} band, and the large frequency shift of the 1,583 cm^{-1} band (to 1,632 cm^{-1}), are indicative that alumina is a strong Lewis acid. On the other hand, the absence of a band at 1,540 cm^{-1} indicates that there are no Brønsted acid sites on the surface strong enough to react with pyridine. The fact that the addition of water vapour did not convert pyridine to the pyridinium ion also indicates that any protons present on the alumina are too weak to react with pyridine.

Alumina treated with hydrogen fluoride[10–12] or boron trifluoride[3] was found to display both greater catalytic activity and increased acidity as measured by either ammonia chemisorption or amine titration.

Comparatively little work has been done on investigating the basic properties of alumina. As mentioned in 3.2.3, Schwab gave evidence for the existence of basic sites on alumina,[13] although some doubt remains over his method for boron trifluoride adsorption.[14] Yamadaya *et al.* have shown by means of benzoic acid titration that basic sites begin to appear on alumina when water sufficient to cover all the acid sites has been adsorbed.[15] Basicities as high as 0.4 mmol/g have been recorded (*cf.* 3.2.1 and Fig. 3–3). Pines *et al.* have carried out a series of interesting stereochemical studies which strongly suggest that alumina contains intrinsic basic sites as well as acid sites, and acts as an acid-base bifunctional catalyst. They used pure alumina catalysts, prepared either by the hydrolysis of aluminium isopropoxide or by the precipitation of aluminium nitrate with ammonia and subsequently calcined at 600 to 800 °C, in investigating the dehydration of menthol, neomenthol, alkylcyclohexanol, 1-decalol, etc.[16]

Acid and basic sites on the surface of alumina can be pictured according to the scheme shown below.[17] The Lewis acid site is visualized as an incompletely coordinated aluminium atom formed by dehydration, and the weak Brønsted site as a Lewis site which has adsorbed moisture, while the basic site is considered to be a negatively charged oxygen atom. The unusual acidity minimum which appears at 600 °C (see Fig. 4–1) is thought to be due to the low crystallinity of η-Al_2O_3 as revealed by X-ray analysis.

$$-O-\underset{}{\overset{OH}{Al}}-O-\overset{OH}{Al}- \xrightarrow[-H_2O]{heat} -O-\underset{\text{Lewis acid site}}{Al^{+}}-O-\underset{\text{Basic site}}{\overset{O^-}{Al}}- \xrightarrow{+H_2O} -O-\underset{\text{Brønsted acid site}}{\overset{H_2O^+}{Al}}-O-\underset{\text{Basic site}}{\overset{O^-}{Al}}-$$

Recently Peri has proposed a detailed scheme for the surface of γ-alumina prepared by heating aluminium hydroxide.[18,19] This scheme, for alumina heat-treated at 800 °C, is shown in Fig. 4–3, where five types of isolated hydroxyl groups labelled A to E are identifiable. Each of them has a different nearest neighbour configuration, and the groups cover 10% of the surface shown (5 out of a possible 49 sites).

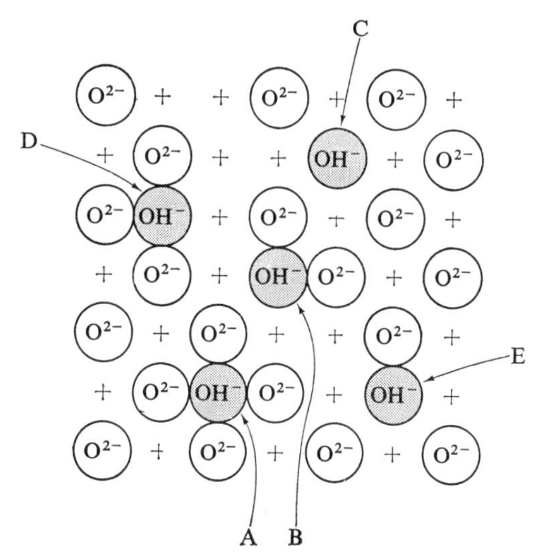

Fig. 4–3 Suggested scheme for acidic and basic sites on γ-alumina
Letters A to E identify the different types of isolated hydroxyl ions, and "+" denotes an Al^{3+} ion on the layer below the surface

The five absorption maxima which are observed in the infrared spectra (at 3,800, 3,780, 3,744, 3,733 and 3,700 cm^{-1}) are associated with sites A, D, B, E and C respectively. Each site has a different local charge density: type A with 4 O^{2-} ions as neighbours is the most negative (a basic site), and type C lacking neighbours is the most positive (an acidic site).[19] Peri also assumes the existence of "acid-base" sites or ion-pair sites on dry γ-alumina on the basis of infrared studies of the adsorption of ammonia.[20] Yamadaya *et al.* attributed the basic property of alumina (which they had detected) to weakly adsorbed free OH groups rather than those OH groups retained on the surface after dehydration, for they noted that the basic property begins to appear when water molecules sufficient to form a mono-molecular layer have been adsorbed, and increases no further after the amount of water corresponds to a layer three molecules thick[15] (*cf.* Fig. 3–3).

The suggestion that the acid sites of alumina which has been dehydrated at high temperature may be formed by lattice distortions has also been put forward by Echigoya[21] and Cornelius *et al.*[22] As the temperature of heat-treatment is raised, a water molecule is removed from two hydroxyl groups attached to aluminium atoms to form an Al–O–Al link (see below). The Al–O–Al link is readily formed between

$$\begin{matrix} H & H \\ | & | \\ O & O \\ | & | \\ Al & Al \end{matrix} \longrightarrow \begin{matrix} & O & \\ / & & \backslash \\ Al & & Al \end{matrix} + H_2O$$

neighbouring Al–OH groups in the early stages of dehydration, but as the distance between them becomes progressively greater with further dehydration, the distortion of the Al–O–Al link (which is responsible for the acid strength) becomes larger.

4.1.2 Alkaline earth metal oxides

Most of the alkaline earth metal oxides show surface basicity. Some acidity is also observed in the case of magnesium oxide calcined at high temperatures as shown in Fig. 3–1. According to the phenol vapour adsorption method (3.1.2), the basic strengths of beryllium oxide and

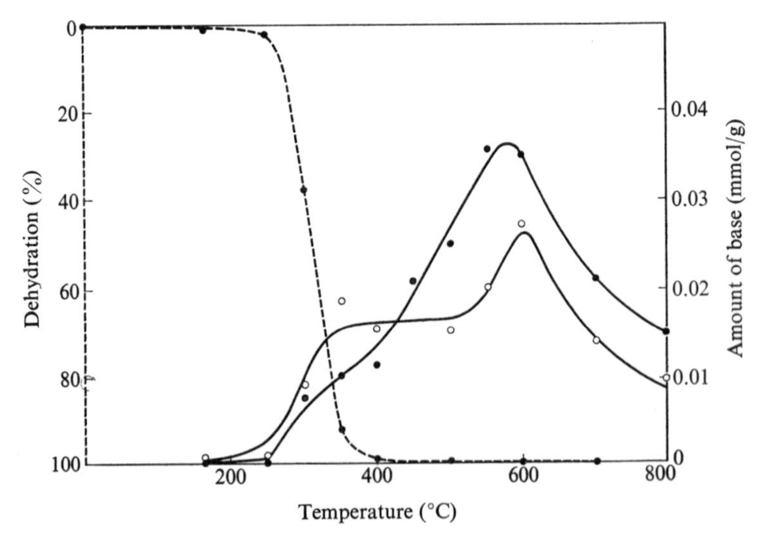

Fig. 4–4 Effect of temperature of heat-treatment on MgO
(—●—) amount of base (indicator; bromothymol blue), (—○—) amount of base (indicator; phenolphthalein), (--●--) percentage dehydration

magnesium oxide may be classified as moderate, and that of calcium oxide as strong.[23] However, bromothymol blue is yellow on beryllium oxide, but blue on four other oxides: MgO, CaO, SrO and BaO. The basicity of calcium oxide as measured by benzoic acid titration using bromothymol blue (3.2.1) has already been given in Fig. 3–2, maximum basicity being observed for calcination at 500 °C. Fig. 4–4 shows the basicity of magnesium oxide measured in the same way using bromothymol blue and phenolphthalein,[24] which is far less than the comparable value for calcium oxide. The calcium oxide in Fig. 3–2 and the magnesium oxide in Fig. 4–4 were both prepared by heating their hydroxides in air for 3 h. The amounts and the strengths of the basic sites on magnesium oxide (prepared by heating its basic carbonate at 600 °C for 20 h) and on calcium oxide (prepared by heating its carbonate at 900 °C for 20 h) are shown in Fig. 4–5, the data for which were obtained by Yoneda

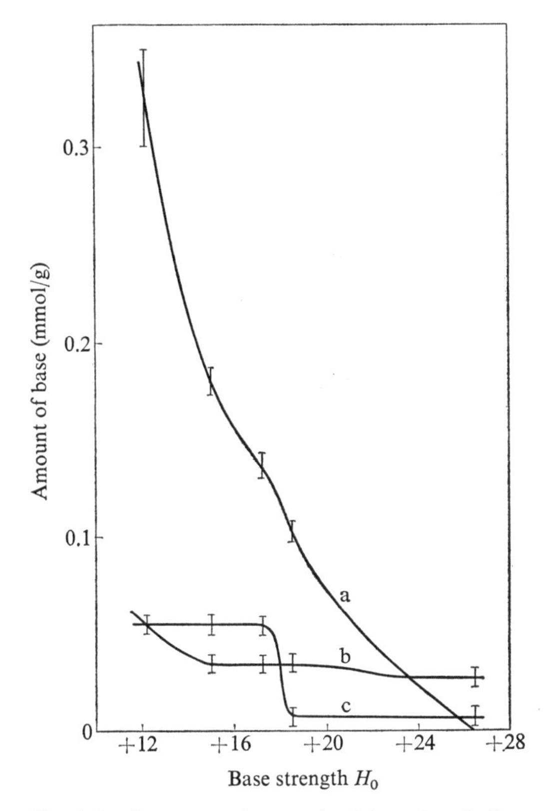

Fig. 4–5 Amount and strength of base for alkaline earth metal oxides
a: MgO b: CaO c: SrO

et al. according to the method described in 3.2.1.[25] It is significant that calcium oxide prepared by heating the hydroxide at 900 °C does not show any basicity even at the weakest basic strength of $H_0 = +7.1$ (Fig. 3–2), whereas a sample prepared from the carbonate shows fairly strong basicity. A similar difference also holds for magnesium oxide: that prepared by heat-treatment of the carbonate at 600 °C is found to have a basicity of about 0.5 mmol/g at $pK_a = +7.1$ (bromothymol blue),[26] a value ten times larger than that for a sample prepared from the hydroxide at 600 °C (see again Fig. 4–4). The strontium oxide sample referred to in Fig. 4–5 was prepared by heating the hydroxide at 850 °C for 20 h. The basicity (the amount of base) at $pK_a = +7.1$ and $+9.3$ for strontium oxide prepared from its carbonate is shown in Fig. 4–6, together with that for barium oxide, also prepared from its carbonate.[27]

The nature or structure of basic sites on alkaline earth metal oxides has not by any means been exhaustively studied, those oxides which have so far been investigated being limited to magnesium and calcium oxide. The presence of several types of basic centres at the surface of partially dehydrated magnesium hydroxide has been discussed by Krylov *et al.* on the basis of studies of adsorption and isotope exchange of carbon dioxide on magnesium oxide and on partially dehydrated magnesium hydroxide surfaces.[28] These centres include 1) strongly

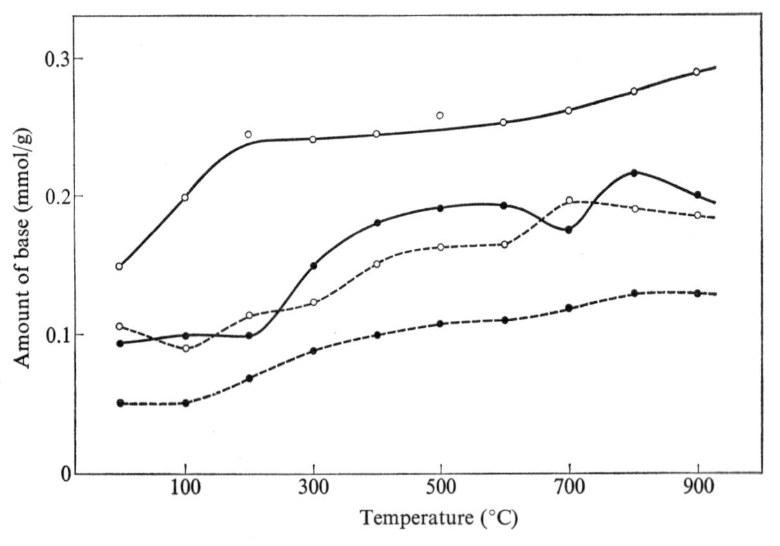

Fig. 4–6 Amount of base *vs.* temperature of heat-treatment for SrO and BaO prepared from their carbonates, at various basic strengths
(—○—) SrO at $pK_a = +7.1$, (—●—) SrO at $pK_a = +9.3$,
(--○--) BaO at $+7.1$, (--●--) at $+9.3$

basic O^{2-} centres which are transformed to CO_3^{2-} ions during the process of adsorption according to the mechanism postulated by Garner, 2) strongly basic centres derived from the O^{2-} ions adjacent to the surface OH groups, and 3) surface OH groups which form weak basic centres. According to the recent work of Malinowski *et al.*, type 3) centres must be ruled out as a possibility on the oxide surface, at least for specimens calcined at temperatures above 600 °C, in which no free OH groups were detected from the infrared spectra.[29)] Recently, the transfer of an electron from magnesium oxide to the adsorbed electron acceptor has been detected by means of ESR and absorption spectra studies.[30-35)] The electron transfer results in the formation of ions and ion radicals which are stabilized at the surface of the adsorbent under vacuum conditions. This is a further indication of the existence of basic sites (in the most general sense) on magnesium oxide. Kortüm found that 1,3,5-trinitrobenzene (I), which is colourless on silica gel, becomes red ($\nu_{max} = 21{,}500\ cm^{-1}$) on magnesium oxide, the red colour disappearing when dissolved out with methanol.[30)] This absorption maximum is the same as that observed in alkaline ethanol solution. Thus it appears that the action of magnesium oxide is analogous to that of OH^- ions as shown below.

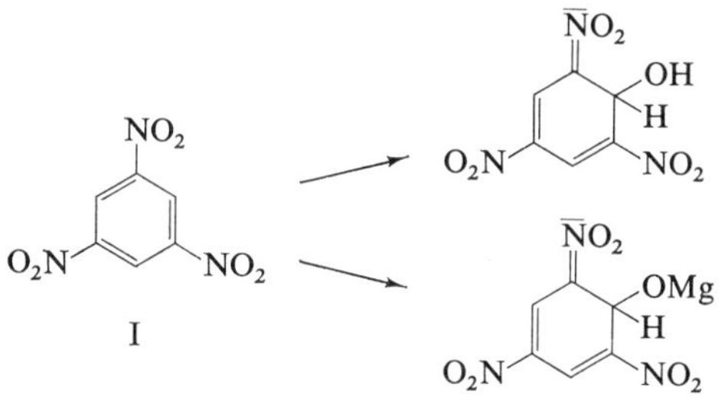

The nature of the basic sites on calcium oxide is considered to be analogous with that of those on magnesium oxide. Here, the change which heat-treatment produces in the IR absorption intensity for the basic OH group which appears at 3,770 cm^{-1}, and the change in ESR signal intensity for the heat-treated sample on which nitrobenzene has been adsorbed[36)] are quite different from the changes in basicity for calcium hydroxide shown in Fig. 3–2. This basicity change is independent of any change in surface area, a point raised again in Chapter 5.

4.1.3 Other oxides and sulfides

Pure silica gel shows neither acidic nor basic properties, but commercial samples generally adsorb methyl red ($pK_a = +4.8$) to give the characteristic red colour, and also sometimes change the basic forms of certain indicators with even lower pK_a values to their acidic forms. Parry has shown from the infrared spectra of adsorbed pyridine that neither Brønsted nor Lewis acidity exist on the surface of silica (Cab–O–Sil from the G.L. Cabot Company), but that hydrogen-bonded pyridine does exist.[9] The infrared spectra of adsorbed ammonia also shows that silica only adsorbs ammonia physically,[37] and has no proton acidity.[38] However, silica which has been treated with ammonium fluoride comes to possess a large population of strong Brønsted acid sites.[39,40] According to Chapman and Hair, the Brønsted acid is formed by weakening the OH bond of a silanol group by the inductive effect of highly electronegative fluorine atoms as illustrated in diagram A.[39] Diagram B shows the precisely analogous mechanism whereby

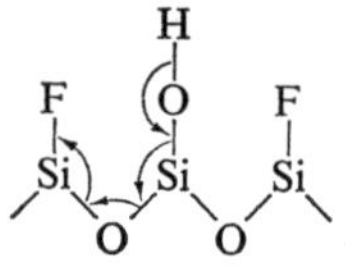

A 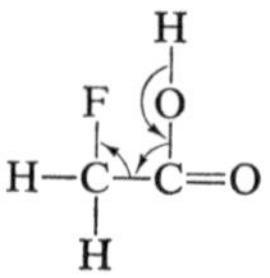 B

the acidity of acetic acid increases when a fluorine atom is substituted for a hydrogen atom of the methyl group.

Ionizing radiation also generates weak acid sites on silica gel.[41] Amine titration using dimethyl yellow has been used to show the subsequent decrease in acidity, the half-time for decay of the acid sites being about 2~3 h at 25 °C and 1~1.5 h at 100 °C.

Commercial zinc oxide is without acid sites stronger than those corresponding to an H_0 value (indicator pK_a) of +4.8, but does display some weak acidity; 0.004 mmol/g at $+4.8 < H_0 \leqq +6.8$. Heat-treatment increases the acidity as shown in Table 4–3.[42] No difference has been detected between the acidities of zinc oxide heat-treated in air or in a vacuum, although it is known that its semiconductivity is affected by changes in the conditions of heat-treatment.[43] On the other hand,

TABLE 4–3 Acidity of ZnO

Temperature of heat-treatment (°C)	Time of heat-treatment (h)	Acid amount at $H_0 \leqq +6.8$ (mmol/g)
300 in air	3	0.010
500 in air	3	0.010
400 in vacuum	3	0.009

zinc oxide which is prepared from the chloride with ammonium hydroxide solution and heat-treated at 300 ~ 400 °C has a large amount of fairly strong acid as shown in Fig. 4–7.[44] According to the infrared study of adsorbed pyridine (2.3.3), the acid sites on the zinc oxide are Lewis type.[44] Although the bromothymol blue indicator test fails to reveal any basic property of zinc oxide,[45] a positive result is obtained by means of the phenol vapour adsorption method, the basic strength being classified as moderate[23] (see 3.1.2). Titanium oxide provides yet another example where the properties of commercial samples and those prepared in the laboratory differ quite markedly. The commercial product displays only very weak acidic properties, but a sample prepared from the chloride and ammonium hydroxide, heat-treated at 400 ~ 500 °C, has fairly strong acid strength as shown in Fig. 4–8.[44]

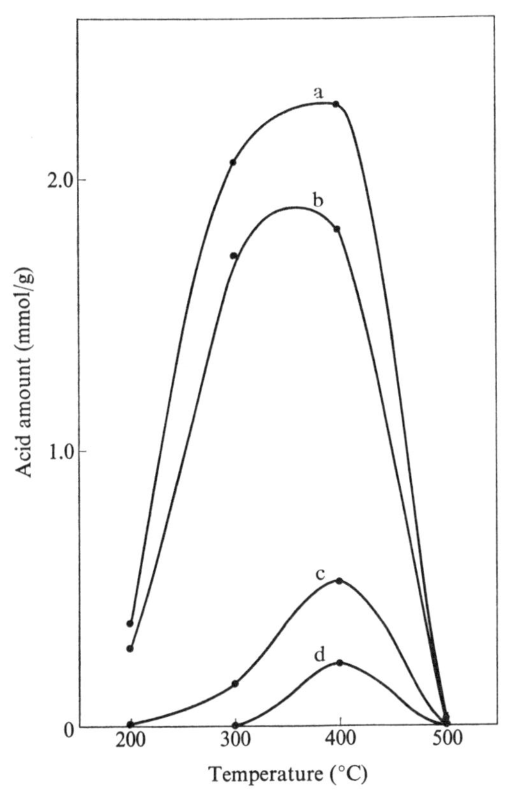

Fig. 4–7 Acid amount for zinc oxide *vs.* heat-treatment temperature at various acid strengths
a: acid strength $H_0 \leqq +6.8$, b: $H_0 \leqq +4.8$,
c: $H_0 \leqq +3.3$, d: $H_0 \leqq +1.5$

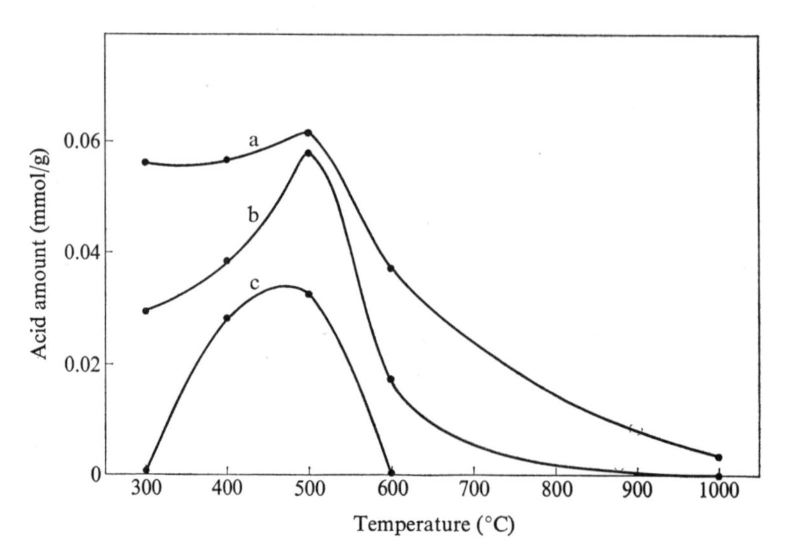

Fig. 4–8 Acid amount for TiO_2 *vs.* heat-treatment temperature at various acid strengths
a: acid strength $H_0 \leqq +4.8$, b: $H_0 \leqq +3.3$, c: $H_0 \leqq +1.5$

Kawaguchi *et al.* report that the acidity of titanium dioxide decreases when the crystal structure changes from that of anatase to that of rutile with heat-treatment at 900 ~ 950 °C.[46] Titanium dioxide has been shown to possess Brønsted type acid sites, unlike those of zinc dioxide, which infrared studies on adsorbed pyridine have revealed to be of Lewis type.[44]

Measurement of the acidic properties of chromic oxide[47] and molybdic oxide[48] using the amine titration method (2.2.2), gives the results shown in Table 4–4 and Fig. 4–9 respectively. The acidity of chromic oxide in its oxidized state is about twice that in the reduced state, and in both cases half (or more) of the total acid sites are strongly acidic. Hirschler *et al.* suggest that the acid sites on oxidized chromic oxide are related to the defects which are also responsible for its semiconducting properties, but the nature of the acid sites on reduced chromic oxide remains obscure.[47] The surface acidity at $H_0 \leqq +4.0$ of MoO_3 changes with the heating temperature as may be seen from Fig. 4–9.

Only qualitative measurements of acidity on V_2O_5, As_2O_3 and CeO_2 have so far been made. V_2O_5, As_2O_3 and heat-treated CeO_2 each produce the characteristic colour changes (yellow to red) in dimethyl yellow and methyl red, but untreated CeO_2 is unable to change the colour of dimethyl yellow.[45,49]

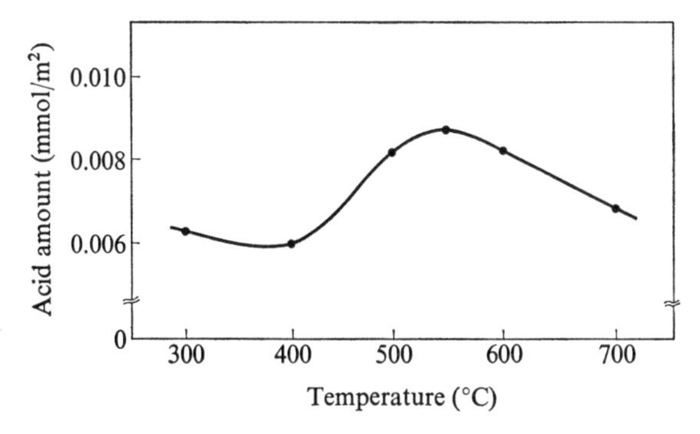

Fig. 4–9 Acid amount at $H_0 \leqq +4.0$ for MoO_3 *vs.* heat-treatment temperature

TABLE 4–4 Acidic properties of Cr_2O_3 †1

Oxidation state†2	Extent of surface oxidation (mmol KI/g)	Acid amount (mmol/g)		
		$H_0 \leqq +3$	$H_0 \leqq -3$	$H_0 \leqq -8$
Oxidized Cr_2O_3	0.46	0.09	0.09	0.05
Reduced Cr_2O_3	0.15	0.04	0.04	0.02

†1 Prepared from chromic nitrate and ammonium hydroxide
†2 Heat-treated with oxygen or hydrogen at 500 °C for 4 h

The acid amount and strength of heat-treated zinc sulfide measured by the amine titration method are shown in Table 4–5. Maximum acidity at $H_0 \leqq +6.8$, $+4.8$ and $+4.0$ is obtained only when heat-treatment is carried out at 300 °C, whereas acid sites at $H_0 \leqq +3.3$ appear only when

TABLE 4–5 Acidity and acid strength of ZnS

Heating temperature (°C)	Acidity (mmol/g)				
	$H_0 \leqq +6.8$	$H_0 \leqq +4.8$	$H_0 \leqq +4.0$	$H_0 \leqq +3.3$	$H_0 \leqq +1.5$
untreated	0.77	0.36	0.092	0	
300	2.8	2.5	1.5	0	
400	†	1.1	0.46	0	
500	1.8	1.2	0.74	0.31	0
650	0.060	0.058	0.01	0	

† Observation of the colour change of neutral red is difficult for this sample

the heat-treatment is at 500 °C.[42] The acidity of the same zinc sulfide has also been measured by the calorimetric titration method using *n*-butylamine (2.2.2), the observed value of 1.3 mmol/g[50] being higher than that found using various indicators with the amine titration method (see Table 4–5). This indicates the effectiveness of the calorimetric method for measuring acid sites weaker than $pK_a = +6.8$. Other metal sulfides whose acidity has been examined are calcium and barium sulfide. Methyl red adsorbed on barium sulfide assumes its yellow colour, but is red on calcium sulfide, while dimethyl yellow remains yellow on both.[49]

4.2 Mixed metal oxides

Silica-alumina, a very well known solid acid, falls into this category. The acidic properties and the nature of the acid sites have been extensively studied employing almost all of the methods described in Chapter 2. Other mixed oxides given in Table 1–1 have not yet been fully investigated, some of them only recently having been found to exhibit acidic properties.

4.2.1 Silica-alumina

Both the acid amount and acid strength for a sample of silica-alumina have already been given in Fig. 2–12. The figure indicates that the sample has very strong acid sites, with strength H_0 of at least -8.2, and amounting to some 0.35 mmol/g. The Brønsted and Lewis acidities of silica-alumina prepared by mixing alumina gel and silica gel and heating at 500 °C are shown in Fig. 4–10.[51] Curve a shows the total acid amount for $H_0 \leqq +1.5$, b the Lewis acid amount measured by the improved Leftin and Hall method using chlorotriphenylmethane (see 2.3.2), c the amount of Brønsted acid estimated from the difference between total amount of acid and Lewis acid alone, and curve d the acid amount measured by the ion exchange method with ammonium acetate (see 2.3.1). Point e gives the chlorotriphenylmethane adsorption at 80 °C on alumina gel. Lewis acidity is a maximum on pure alumina gel, decreasing steadily with increasing silica gel content. Brønsted acidity, on the other hand, reaches its maximum when the proportion of silica reaches 70%. Silica-alumina prepared in this way shows less Lewis acidity and more Brønsted acidity than a mixed gel which has

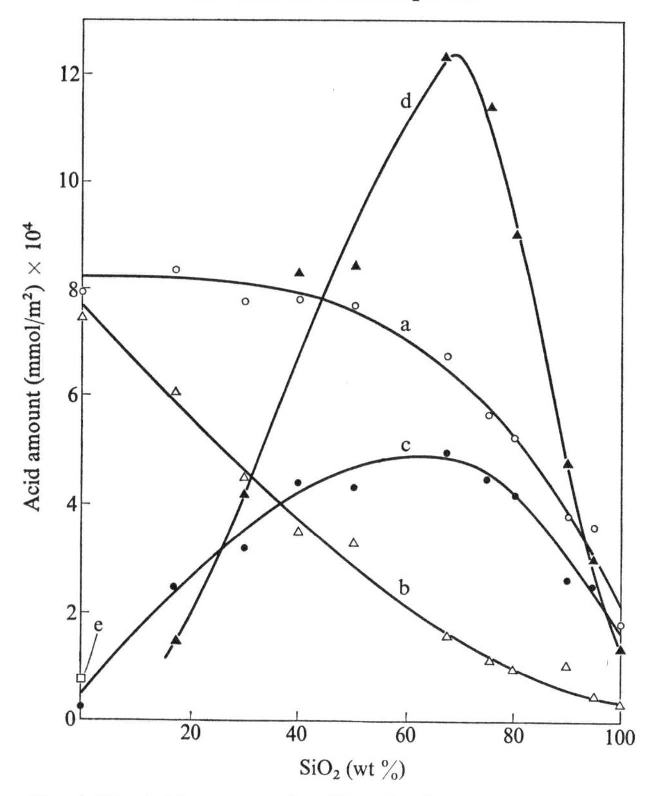

Fig. 4–10 Acid amounts for silica-alumina *vs.* proportion of silica (wt%)
a (○) total amount of acid at $H_0 \leqq +1.5$
b (△) Lewis acid amount
c (●) Brønsted acid amount (c = a−b)
d (▲) amount of acid determined by ion exchange
e (□) chlorotriphenylmethane adsorption at 80 °C on 100% alumina

been co-precipitated and has co-gelated.[51] The Brønsted acidity measured by the above method is not in accord with that obtained by the ion exchange method, which is specific to Brønsted acids. This is considered to be due largely to the influence of water molecules in the latter method, and at least partly due to the fact that curve c was derived from the difference between the acidity for fairly strong acid ($H_0 \leqq +1.5$) and the Lewis acidity as measured by a method considered unreliable for the reasons given in 2.3.2. An unambiguous differentiation between Lewis and Brønsted acidity is, however, possible through the use of infrared techniques. Recent studies of the infrared spectra of pyridine chemisorbed on synthetic silica-alumina have indicated the presence of both Lewis and Brønsted acid sites.[9,52,53] The infrared spectra of pyridine adsorbed on silica-alumina (Houdry M–46 cracking catalyst) are shown in Fig. 4–11,[9] where both the pyridinium ion (at 1,540 cm^{-1}) and strongly coordinately bonded pyridine (at 1,449 cm^{-1}) may be discerned on curve a (*cf.* 2.3.3 and Table 2–6). Spectrum b, obtained after evacuation at 300 °C, still indicates the presence of some pyridinium ions, but clearly shows the presence of Lewis acidity on the crack-

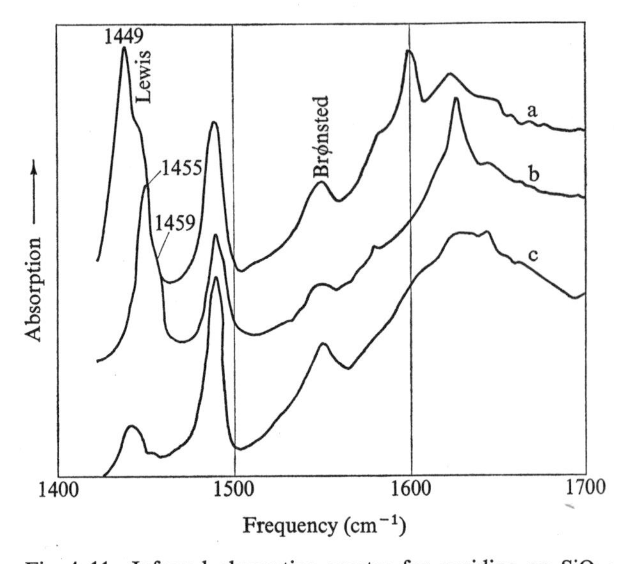

Fig. 4–11 Infrared absorption spectra for pyridine on $SiO_2 \cdot Al_2O_3$
a: equilibriated with pyridine, evacuated at room temperature, b: after evacuation at 300 °C, c: after adding 0.05 mmol H_2O

ing catalyst which would appear to be quite marked (1,455 and 1,459 cm^{-1} bands). Spectrum c shows the changes which occur upon the addition of 0.05 mol of water to the sample. The increase in the 1,540 cm^{-1} band indicates that a considerable amount of Brønsted acid has been formed, and the decrease in the 1,450 cm^{-1} band shows a concomitant decrease in Lewis acidity. This conversion of Lewis sites to Brønsted sites by water molecules has been confirmed by titration[54] and from studies of the infrared spectrum of adsorbed ammonia.[55] The latter consisted of a quantitative study of the ratio of Lewis to Brønsted acid sites on silica-alumina (see 2.3.3 and Table 2–5). The results revealed that the only detectable adsorbed species were physically adsorbed NH_3 [PNH_3], coordinately bonded NH_3 [LNH_3], and NH_4^+. From the relative intensities of the appropriate bands, the ratio of $[NH_3] = [PNH_3] + [LNH_3]$ to $[NH_4^+]$ was found to be 4 to 1 at low concentrations, as shown in Fig. 4–12. Hall *et al.* suggest, on the basis of the deuterium exchange experiment and observation of the NMR spectra, that the hydrogen on the surface of silica-alumina is chemically similar

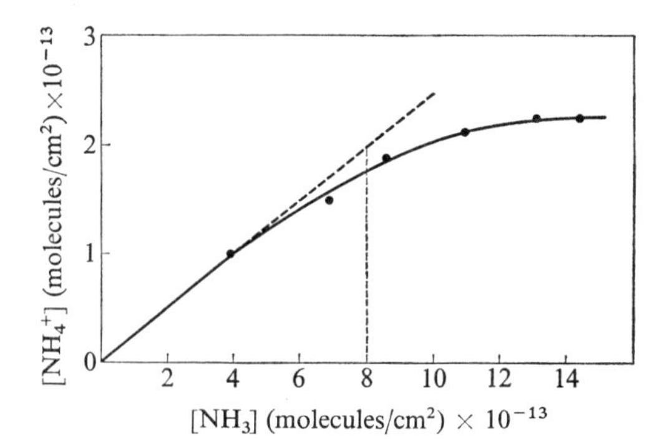

Fig. 4–12 Population of $[NH_4^+]$ *vs.* $[NH_3]$, giving the ratio of Lewis to Brønsted acidity on $SiO_2 \cdot Al_2O_3$

to that in alcohol, most of the hydrogen atoms existing in the form of SiOH, some of them forming AlOH, with an upper limit on the number of Brønsted acid sites of $3 \times 10^3 H^+/cm^2$.[56] With the total acidity (B+L) of silica-alumina revealed by amine titration to be generally about 0.5 mmol/g, calculations set the amount of Brønsted acid at about 25% of the total amount, assuming a surface area for the sample of 250 m²/g. This value is in good agreement with that obtained from the infrared spectra of adsorbed ammonia. However, Basila *et al.* have concluded, from infrared spectroscopic studies of pyridine chemisorbed on a typical silica-alumina cracking catalyst, that there are essentially equal numbers of Lewis and Brønsted acid sites present on the surface.[52] There are some other data on the acidic properties of silica-alumina which will be given in connection with catalytic activity in the following chapter. It should perhaps be pointed out at this stage that the acidic properties of silica-alumina are to a large extent dependent upon the method of preparation, the proportion of alumina, the temperature of dehydration, and the method by which the acidity is assessed. Changes in these acidic properties due to both grinding[57] and steam-heating[58] have also been reported.

The adsorption of ammonia on silica-alumina has only a minor effect on the total number of sites, but a pronounced effect on the acid strength distribution.[6] Also there is the fact that just as silica treated with NH_4F (*cf.* 4.1.3) has increased surface acidity, so treatment of silica-alumina with a compound containing fluorine enhances its catalytic activity.[59] It is already well-established that the catalytic activity of silica-alumina in cumene cracking[60,61] and in the double-bond isomerization of butene[62] is enhanced by prior irradiation with γ-rays or neutrons. Mitsutani *et al.* have found that preirradiation with γ-rays or electrons increases the acidity of silica-alumina as measured by the amine titration method with dimethyl yellow ($pK_a = +3.3$) as shown in Fig. 4–13, which shows the close correlation between surface acidity and propylene polymerization activity.[63] A similar phenomenon has been observed in the dehydration of alcohols.

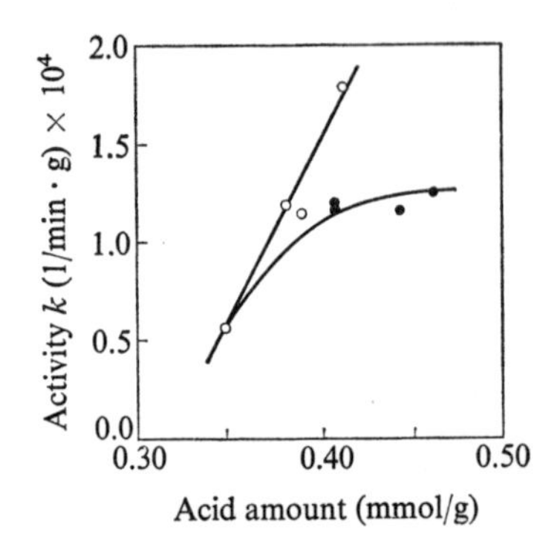

Fig. 4–13 Propylene polymerization activity k *vs.* acid amount at $pK_a = +3.3$ on irradiated silica-alumina surface
(○) irradiated with electrons,
(●) irradiated with γ-rays

The basic properties of silica-alumina have also been studied recently by calorimetric titration with trichloroacetic acid[50] (*cf.* 3.2.4). The titration curve, from which the amount of base on the sample was calculated to be *ca.* 0.6 mmol/g, has already been given in Fig. 3–6. Since trichloroacetic acid is a fairly strong acid, the observed basicity must include relatively weak basic sites. Nevertheless, the fact that basic sites as well as acid sites do exist on silica-alumina is of great importance from the view-point of acid-base bifunctional catalysis.

The structure of acid centres on silica-alumina has been discussed in detail by a number of workers including Hansford,[64] Yamaguchi and Tsutsumi,[65] Thomas,[66] Tamele,[67] Milliken *et al.*,[68] Danforth,[69] Plank,[70] Greensfelder *et al.*,[71] Oblad *et al.*,[72] Emmett and Haldeman,[73] Mills and Hindin,[74] Lee and Weller,[75] Basila *et al.*,[52] Fripiat *et al.*,[76] Hall *et al.*,[77] Peri[78] and Hirschler.[6] Their conclusions are based on studies of the physical properties, catalytic activity, cation exchange rate, the exchange reaction with D_2O, D, $H_2{}^{18}O$ or ${}^{18}O_2$, and electron diffraction. It is generally believed that the acid centres, whether of Brønsted or Lewis type, owe their existence in silica-alumina to an isomorphous substitution of trivalent aluminium for tetravalent silicon in the silica lattice.[64,66,67] Isomorphous substitution of this kind would lead to a structure like that given below in diagram A. Because

A $\quad -\mathrm{Si}-\mathrm{O}-\mathrm{Al}^{-}-\mathrm{O}-\mathrm{Si}-$

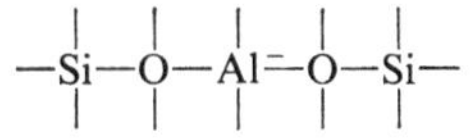

B $\quad -\mathrm{Si}:\ddot{\mathrm{O}}:\leftarrow\mathrm{Al}\rightarrow:\ddot{\mathrm{O}}:\mathrm{Si}-$ (with $\downarrow :\ddot{\mathrm{O}}: -\mathrm{Si}-$ below Al) $\qquad -\mathrm{Si}:\ddot{\mathrm{O}}:\leftarrow\mathrm{Al}\rightarrow:\ddot{\mathrm{O}}:\mathrm{Si}-$ (with $\mathrm{H}:\ddot{\mathrm{O}}:\mathrm{H}^{+}$ above Al and $\downarrow :\ddot{\mathrm{O}}: -\mathrm{Si}-$ below Al)

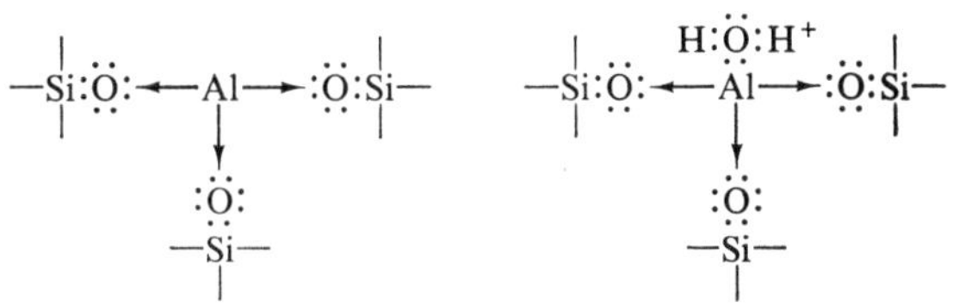

Lewis acid site $\qquad$ Brønsted acid site

the normally six-coordinated aluminium atom has been forced to assume a four-coordinated structure, a net unit negative charge is created at this point on the solid surface, requiring neutralization by a cation such as a proton.[66] A similar view, presented by Tamele[67] and Hansford,[64] is that the aluminium atom under these conditions tends to acquire a pair of electrons to fill its p-shell, creating a Lewis acid in the absence of water, and a Brønsted acid in the presence of one water molecule (see diagram B). On the basis of a number of experimental observations, Basila *et al.* suggest that all the primary acid on a typical silica-alumina sample is of the Lewis type, centred on active surface aluminium atoms, and that each apparent Brønsted site is produced by second-order interaction between a molecule (e.g. H_2O, etc.) chemisorbed on a primary Lewis acid site and a nearby surface OH group.[52]

The approximate average distribution of surface groups on silica-alumina is given in Table 4–6 from data compiled by Basila *et al.*[52] The data indicate that there are roughly two surface aluminium atoms and five or six surface silicon atoms per surface OH group. It is also evident that only a small fraction of oxygen atoms on the surface are in the form of surface OH groups; similarly, only a small proportion of the surface aluminium atoms are located in acidic sites strong enough to chemisorb pyridine at 150 °C.

We now proceed to the discussion of the detailed consequences

TABLE 4–6 Average distribution of surface groups on $SiO_2 \cdot Al_2O_3$ †1

(OH)s/Als = 0.55†2
Os/Als = 6.62
Sis/Als = 2.50
(OH)s/Os = 0.084
K/Als ≧ 0.6†3
C_5H_5N/Als = 0.087†4

†1 Subscript s indicates a surface species
†2 Based on a value of 1.24×10^{14} surface OH/cm^2 for American Cyanamid Aerocat Triple A $SiO_2 \cdot Al_2O_3$
†3 For the base (potassium acetate) exchanged $SiO_2 \cdot Al_2O_3$, 0.21 gram-atom of K per gram-atom of Al
†4 Chemisorbed at 150 °C

of site structure. According to Hall *et al.*, the extent to which ammonia is chemisorbed per unit area of silica-alumina in the range of temperatures 175 ~ 500 °C does not appear to be a function of catalyst hydrogen content as measured by deuterium exchange.[77] Upon extensive dehydration, even at 500 °C, an amount of ammonia several times greater than that of the residual hydrogen was still adsorbed. This strongly suggests that the strong acid sites which are presumed to adsorb ammonia chemically are not of the Brønsted type. Certain sites on dry silica-alumina have been shown to be characterized by strong physical adsorption of carbon dioxide (as revealed by the presence of its characteristic absorption band at 2,375 cm^{-1}).[78] These "α-sites" exist on silica-alumina at concentrations of less than 10^{13} sites/cm^2. Acetylene, butene, HCl and various other compounds are adsorbed with very high selectivity on these sites. NH_3 and H_2O are adsorbed far less selectively. Peri has suggested that such sites include an O^{2-} ion, and are "acid-base" rather than simply Lewis acid sites.[78] Shirasaki *et al.* have investigated the nature of the OH groups on silica-alumina (Nalcat, 13.5 wt% Al_2O_3) and cation-exchanged catalysts (Na^+, K^+, Ni^{2+}, Co^{2+}, Cr^{3+}, etc).[79] Their results both from infrared techniques and from the measurement of acidity suggest that Si–OH, mostly concentrated at the surface, forms Brønsted acid sites (see Fig. 4–14, where Brønsted acid is marked as $\overline{H}$). This acid can, of course, undergo cation exchange in aqueous solutions without, it is thought, affecting the surface silica-alumina hydrogen bonds. The latter, however, are broken when the catalyst is dispersed in an aprotic solvent, with a reduction in the acidity due to the increased difficulty in liberating the $\overline{H}$ protons. It was also found that the rate at which the hydrogen bonds break is slower in silica-alumina than in the cation-exchanged catalysts.

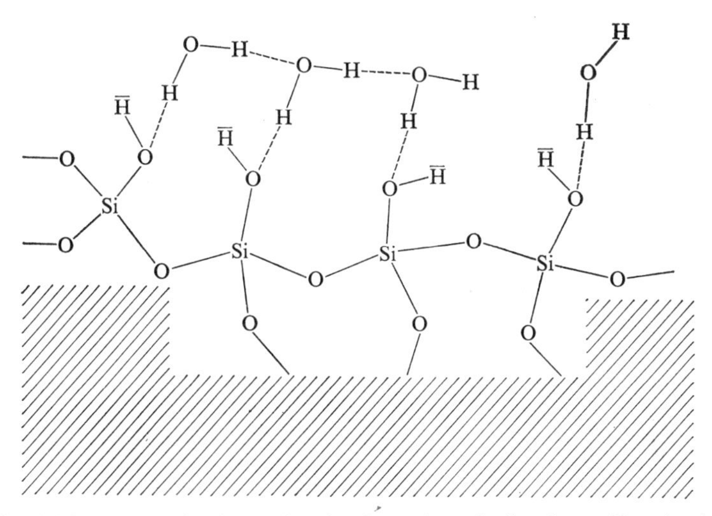

Fig. 4–14 Suggested scheme for the formation of silanol on silica-alumina surfaces
$\overline{\mathrm{H}}$: protonizable hydrogen of silanol (Brønsted acid site)

Hirschler, in order to explain his observation that ammonia adsorption decreases acid strength without decreasing the total number of acid sites, and that one molecule of a nitrogen base eliminates two strong acid sites, has proposed a model of the acid site.[6] The model, given below, would explain the above observations if, as would be expected, both of the Si–OH groups pictured are acidic.

Plank, however, has proposed an alternative explanation for the

```
HO    O          HO    O
  \  / \           \  /
   Si   Al          Si
  / \  / \         / \
 O   OO   O       O   O
```

Strong Brønsted acid

```
HO    O Na⁺ OH   HO    O
  \  / \  /        \  /
   Si   Al⁻         Si
  / \  / \         / \
 O   OO   O       O   O
```

Base-exchanged site, weaker Brønsted acid

```
HO    O Na⁺ O    O
  \  / \  /  \  /
   Si   Al⁻   Si
  / \  / \   / \
 O   OO   OO    O
```

Base-exchanged and dehydrated

acidity of silica-alumina.[70] From a study of the differences between silica and silica-alumina gels, he concluded that the Al_2O_3 molecule always becomes a terminal group in the micelle structure, and that therefore isomorphous substitution of aluminium for silicon does not occur. According to his hypothesis, aluminium ions in the terminal alumina groups are coordinated with OH groups and water in such a way as to retain their normal hexagonal coordination as illustrated in the diagram below. Application of Pauling's electrostatic valence

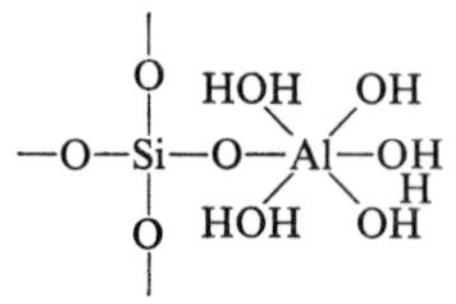

```
       |
       O  HOH   OH
       |     \ /
   —O—Si—O—Al—OH
       |     / \  H
       O  HOH   OH
       |
```

rule, that is distributing the trivalent positive ion of aluminium among the six oxygen-aluminium bonds, would result in lability of the hydrogen atoms in the coordinated water molecules, that is the creation of Brønsted acid sites.[64]

4.2.2 Alumina-boria, silica-zirconia, and silica-alkaline earth oxides

The acidic properties of alumina-boria ($Al_2O_3 \cdot B_2O_3$) have recently been measured by Izumi and Shiba.[80] As shown in Fig. 4–15, the acid sites are almost exclusively of the Brønsted type, with the acidity at its maximum when the concentration of B_2O_3 is 15 wt%. Fig. 4–15 shows the results for η-$Al_2O_3 \cdot B_2O_3$. Those for χ-$Al_2O_3 \cdot B_2O_3$ are almost identical. The acidity (acid amount) at $H_0 \leqq +1.5$ was measured by amine titration, the Lewis acidity by the improved Leftin and Hall method (see 2.3.2), and the Brønsted acidity was obtained by subtracting the Lewis acidity from the total (B+L) acidity at $H_0 \leqq +1.5$ (*cf.* 4.2.1).

Silica-zirconia ($SiO_2 \cdot ZrO_2$) shows very strong acidity, with $H_0 \leqq -8.2$ even where the ZrO_2 content is only 0.1%.[81] The change in the acid amount with change of ZrO_2 content will be cited in the following chapter in connection with the corresponding catalytic activity. It has also been established that alumina-zirconia in the proportion of 90% alumina to 10% zirconia adsorbs dicinnamalacetone ($pK_a = -3$) to give the characteristic red colour.

Silica-magnesia ($SiO_2 \cdot MgO$), on the other hand, possesses little acidity when the magnesia content is below 10%, but exhibits modera-

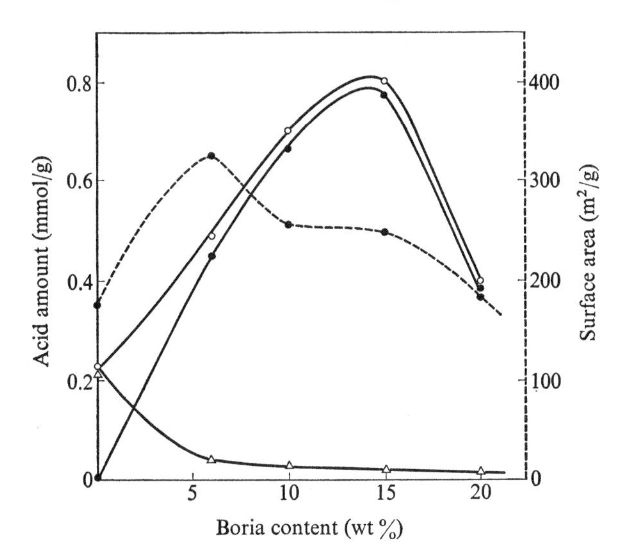

Fig. 4–15 Acid amount for alumina-boria *vs.* boria content (wt%) with corresponding surface area; sample prepared from η-alumina
(—○—) total amount of acid at $H_0 \leqq +1.5$,
(—△—) Lewis acid amount,
(—●—) Brønsted acid amount,
(--●--) surface area

tely strong acidic properties at higher concentrations.[81] According to Benesi's measurements, the acid amount can reach as much as 0.8 mmol/g in the range of H_0 from $+2$ to -3, as has already been illustrated in Fig. 2–12. Table 4–7 shows the effect of the calcinating temperature on the acidity of silica-magnesia prepared by hydrolyzing an equimolar mixture of silicon tetraethoxide and magnesium ethoxide.[82] The silica-magnesia calcined at lower temperatures has a relative

TABLE 4–7 Acidic properties of $SiO_2 \cdot MgO$

Calcinating temperature (°C)	Acidity (μmol/m²)		
	$H_0 = +0.8 \sim +3.3$	$H_0 = +3.3 \sim +5.1$	$H_0 = +5.1 \sim +6.8$
400	1.73	0.23	0.06
500	1.43	0.74	0.13
600	1.70	0.92	0.42
700	1.52	0.85	1.02
800	0.28	1.19	2.48

abundance of moderately strong acid sites and a small number of weak sites, whereas in a sample calcined at 800 °C the number of moderately strong acid sites is small, and the number of weaker sites proportionately larger. Although the high acid strength reported by Dzisko *et al.*[83] for a similar sample was not reproduced by Bremer and Steinberg[82] (see again Table 4–7), Echigoya *et al.* have recently found the high acid strengths and high basicities given in Fig. 4–16 and Table 4–8.[84] Their sample was prepared from silica-magnesia which had been obtained by mixing colloidal silica with $Mg(OH)_2$ and heating it to 290 °C under a pressure of 80 ∼ 90 kg/cm^2 ammonia for 24 h. Table 4–8 gives the acidic and basic properties together with the surface area and the structure of other silica-alkaline earth metal oxides prepared by the same method. The order of the acid strength as well as the amount per unit surface area was found to be $SiO_2 \cdot MgO > SiO_2 \cdot CaO > SiO_2 \cdot SrO > SiO_2 \cdot BaO$.

Infrared investigations of the nature of the acid centres on $SiO_2 \cdot MgO$ indicate the presence of Lewis acid sites and weakly acidic isolated OH groups.[82] In the case of $SiO_2 \cdot ZrO_2$, a model has been presented by Thomas as shown below.[66] This requires that zirconium should be

```
 [       |      |      |         ]4-
 [     —Si    —Si—    Si—        ]
 [       | \    |    / |         ]
 [          O   O   O            ]
 [       |   \  |  /     |       ]
 [ —Si—O——————Zr——————O—Si—      ]   4H+
 [    |      /  |  \     |       ]
 [          O   O   O            ]
 [       | /    |    \ |         ]
 [     —Si    —Si—    Si—        ]
 [       |      |      |         ]
```

considered within an eight-coordinated active structure. Such a structure requires four positive ions (protons) for electrostatic neutrality. However, it would seem that the structure might be too unstable to permit the preparation of an active $SiO_2 \cdot ZrO_2$ catalyst suitable for high-temperature reactions. A six-coordinated structure with a net negative charge of two appears more likely.[64] A structure similar to that of Plank's model for $SiO_2 \cdot Al_2O_3$ on page 66, involving a six-coordinated zirconium atom in $SiO_2 \cdot ZrO_2$ or a four-coordinated magnesium atom in $SiO_2 \cdot MgO$, would also represent a relatively strong Brønsted acid.

The structure of alumina-boria has been studied by X-rays, DTA, infrared spectra and acidity measurement.[80] The dehydration of boric acid gives effective acid centres in active alumina, the reaction with alumina taking place quite freely at 130 °C, through the probable intermediary of metaboric acid. Most of the acid sites are of the Brønsted

TABLE 4–8 Acidic and basic properties of SiO_2–alkaline earth metal oxides

	Treatment	Surface area (m^2/g)	pK_a of surface	Acidity†[1]		Basicity†[2]		Structure
				mmol/g	10^{-4} mmol/m^2	mmol/g	10^{-4} mmol/m^2	
$SiO_2 \cdot MgO$	untreated	125	−1.5	0.203	16.2	0.50	40	Amorphous + [$Mg(OH)_2$]
$SiO_2 \cdot MgO$	290 °C, 24 h	194	−5.6	0.433	22.4	—	—	Sepiolite
$SiO_2 \cdot CaO$	untreated	17.9	+6.8	0	0	0.065	36.3	Amorphous
$SiO_2 \cdot CaO$	290 °C, 24 h	45.7	+3.3	0.034	7.54	0.063	13.7	Xonotlite
$SiO_2 \cdot SrO$	untreated	34.2	—	0	0	0.054	15.8	Amorphous
$SiO_2 \cdot SrO$	290 °C, 24 h	9.02	+4.8	0.003	<3.3	0.017	18.9	Crystalline†[3]
$SiO_2 \cdot BaO$	untreated	4.65	—	0	0	—	—	Amorphous
$SiO_2 \cdot BaO$	290 °C, 24 h	1.98	+4.8	0	0	0.006	30.3	Crystalline†[4]

†[1] Measured by amine titration method using dimethyl yellow ($pK_a = +3.3$)
†[2] Measured by benzoic acid titration using bromothymol blue ($pK_a = +7.1$)
†[3] Subsequently identified as strontium metasilicate
†[4] Subsequently identified as barium metasilicate

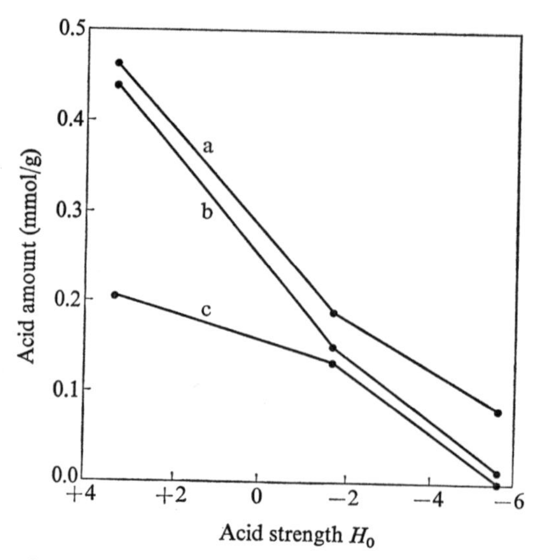

Fig. 4–16 Acid amount *vs.* acid strength for silica-magnesia heat-treated at 290 °C for a: 24 h, b: 5 h, and c: untreated[84]

type as may be seen from Fig. 4–15. Alumina-boria does not show any definite chemical structure like that which characterizes aluminium borates such as $9Al_2O_3 \cdot B_2O_3$, $2Al_2O_3 \cdot B_2O_3$ or $Al_2O_3 \cdot B_2O_3$. It has been suggested that linkages of the type Al–O–B–O–Al may be formed between points on the surfaces of the alumina crystal structure, with the formation of acidic centres.

4.2.3 Other mixed metal oxides

The acid strengths of $Ga_2O_3 \cdot SiO_2$, $BeO \cdot SiO_2$, $Y_2O_3 \cdot SiO_2$, $La_2O_3 \cdot SiO_2$, $SnO \cdot SiO_2$ and $PbO \cdot SiO_2$ have been determined by the spectrophotometric method using various basic indicators listed in Table 2–3 (see 2.1.2).[85] The results are given in Table 4–9, together with those of $Al_2O_3 \cdot SiO_2$, $ZrO_2 \cdot SiO_2$ and $MgO \cdot SiO_2$ for comparison.

The acidic properties of $Cr_2O_3 \cdot Al_2O_3$ are shown in Table 4–10.[47] The acid amount per gram of this catalyst exceeds that found for Cr_2O_3 (*cf.* Table 4–4). Also $Cr_2O_3 \cdot Al_2O_3$ suffers a reduction in its ability to chemisorb quinoline after the addition of alkali.[86] The amounts of quinoline adsorbed by this kind of catalyst are in any case less than the amount of *n*-butylamine taken up from benzene solution. The difference

TABLE 4–9 Acid strengths of some mixed oxides

Mixed oxides (mol%)	Surface area (m^2/g)	Acid strength H_0
$Al_2O_3 \cdot SiO_2$ (94 : 6)†1	270	−8.2
$ZrO_2 \cdot SiO_2$ (88 : 12)†2	440	−8.2∼−7.2
$Ga_2O_3 \cdot SiO_2$ (92.5 : 7.5)†1	90	−8.2∼−7.2
$BeO \cdot SiO_2$ (85 : 15)†1	110	−6.4
$MgO \cdot SiO_2$ (70 : 30)†3	450	−6.4
$Y_2O_3 \cdot SiO_2$ (92.5 : 7.5)†1	110	−5.6
$La_2O_3 \cdot SiO_2$ (92.5 : 7.5)†1	80	−5.6∼−3.2
$SnO \cdot SiO_2$ (85 : 15)†1	70	weaker than $La_2O_3 \cdot SiO_2$
$PbO \cdot SiO_2$ (85 : 15)†1	100	weaker than $La_2O_3 \cdot SiO_2$

†1 Respective metal hydroxides were precipitated on dispersed SiO_2
†2 Ethyl silicate and zirconium nitrate were coprecipitated
†3 Ethyl silicate and magnesium acetate were coprecipitated

TABLE 4–10 Acidic properties of $Cr_2O_3 \cdot Al_2O_3$†1

Oxidation state†2	Extent of surface oxidation (mmol KI/g)	Acid amount (mmol/g)		
		$H_0 \leqq +3$	$H_0 \leqq -3$	$H_0 \leqq -8$
Oxidized $Cr_2O_3 \cdot Al_2O_3$	0.68	0.25	0.25	0.19
Reduced $Cr_2O_3 \cdot Al_2O_3$	0.08	0.16	0.16	0.10
Oxidized $Cr_2O_3 \cdot Al_2O_3 \cdot K_2O$†3	0.86	0.14	0.14	0.01
Reduced $Cr_2O_3 \cdot Al_2O_3 \cdot K_2O$†3	0.04	0.07	0.07	0.02

†1 Houdry $Cr_2O_3 \cdot Al_2O_3$ catalyst, series A, grade 100
†2 Heat-treated with oxygen or hydrogen at 500°C for 4 h
†3 Houdry $Cr_2O_3 \cdot Al_2O_3$ catalyst impregnated with 1% K_2O

is probably due to the large size of the quinoline molecule or its relatively low basicity. With reduced $Cr_2O_3 \cdot Al_2O_3$, the number of ethylenediamine equivalents required to neutralize the acidity is exactly twice that for *n*-butylamine. This result can be understood as implying that the acid sites are too far apart on the surface for a single diamine molecule to react with two sites. In the case of oxidized $Cr_2O_3 \cdot Al_2O_3$, the butylamine titre exceeds that required for the reduced catalyst by 0.09 mmol/g, while the ethylenediamine titre is also higher by 0.13 mmol/g. This result would indicate that for 64% of the chromia acid sites an ethylenediamine molecule is able to neutralize two chromia acid sites (or one alumina site plus one chromia acid site). It might also indicate that the distance between two chromia acid sites or between a chromia and an alumina acid site is less than the distance between two alumina sites.[47]

It has been recently found that $MoO_3 \cdot Fe_2(MoO_4)_3$ shows some acidity, which has a close correlation with its catalytic activity in the oxidation of methanol to formaldehyde (see Fig. 4–17).[48] Pure $Fe_2(MoO_4)_3$ has no acidity or measurable activity, but $MoO_3 \cdot Fe_2(MoO_4)_3$ shows acidity which exceeds that of MoO_3 alone (*cf.* Fig. 4–9).

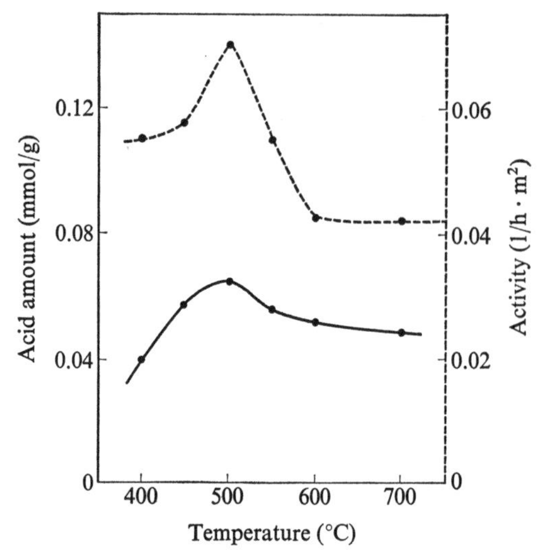

Fig. 4–17 Effect of heat-treatment on $MoO_3 \cdot Fe_2(MoO_4)_3$, wt ratio 2.5/1
(--●--) catalytic activity,
(—●—) amount of surface acid at $H_0 \leqq +4.0$

$TiO_2 \cdot ZnO$ prepared by coprecipitating the respective chlorides with ammonium hydroxide is found to possess a large amount of fairly strong acid, as shown in Fig. 4–18.[44] The acid is stronger than that of either ZnO or TiO_2 alone (see Fig. 4–7 and Fig. 4–8). As mentioned in 4.1.3, the acid sites of ZnO and TiO_2 are of Lewis and Brønsted type respectively. The infrared spectra of adsorbed pyridine indicate that $TiO_2 \cdot ZnO$ possesses both Brønsted and Lewis acid sites. It is interesting to note that even for a sample where the proportion of TiO_2 to ZnO is 19:1, the number of Lewis acid sites is comparable with that of Brønsted sites.[44]

The following mixed oxides show catalytic activity in the hydration of propylene, and are therefore presumed to possess some fairly strong

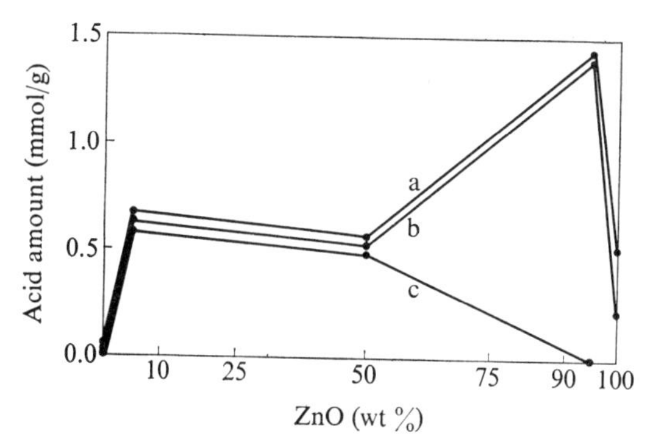

Fig. 4–18 Acid amount for $TiO_2 \cdot ZnO$ *vs.* proportion of ZnO (wt%) at various acid strengths
a: $H_0 \leqq +3.3$, b: $H_0 \leqq +1.5$, c: $H_0 \leqq -3$

acid properties, although none in fact have as yet been measured.[87] 1) ZnO plus other oxides ($ZnO \cdot ZrO_2$, $ZnO \cdot Nb_2O_5$, $ZnO \cdot SnO_2$, $ZnO \cdot Sb_2O_3$, $ZnO \cdot Tl_2O_3$, $ZnO \cdot Bi_2O_5$), 2) CdO plus other oxides ($CdO \cdot Tl_2O_3$, $CdO \cdot GeO_2$, $CdO \cdot ThO_2$, $CdO \cdot Sb_2O_3$, $CdO \cdot TeO_2$), 3) TiO_2 ($TiO_2 \cdot V_2O_5$, $TiO_2 \cdot As_2O_5$, $TiO_2 \cdot Sb_2O_5$, $TiO_2 \cdot MoO_3$, $TiO_2 \cdot WO_3$, $TiO_2 \cdot TeO_3$), 4) HgO ($HgO \cdot Tl_2O_3$, $HgO \cdot CeO_2$, $HgO \cdot ThO_2$), 5) SnO_2 ($SnO_2 \cdot As_2O_5$, $SnO_2 \cdot Sb_2O_5$, $SnO_2 \cdot TeO_3$, $SnO_2 \cdot Bi_2O_5$), 6) MnO_2 ($MnO_2 \cdot V_2O_5$, $MnO_2 \cdot As_2O_5$), 7) Fe_2O_3 ($Fe_2O_3 \cdot TiO_2$, $Fe_2O_3 \cdot MoO_3$, $Fe_2O_3 \cdot WO_3$, $Fe_2O_3 \cdot SnO_2$, $Fe_2O_3 \cdot Sb_2O_5$).

4.3 Natural clays (zeolites, etc.)

The natural clays which are included in Table 1–1 mainly consist of oxides of silicon and aluminium. The acid strengths of some of these clays have already been given in Table 2–2. The strength and amount of acid for kaolinite, attapulgite and montmorillonite, which were measured by Benesi's *n*-butylamine method,[88] are shown in Fig. 4–19. Both the strength and the amount are less than that of synthesized $SiO_2 \cdot Al_2O_3$ samples (*cf.* Fig. 2–12). Although these clays have been used as catalysts for many years, little work has been done to determine their acidic properties and the nature of the acid sites. However, increasing attention is being paid to zeolites, and in particular to synthesized zeolites. Zeolites are a class of hydrated aluminium silicates, that is the metal salts of hydrated alumino-silicic acid, consisting of a three dimensional crystalline network of –Al–O–Si– atoms in the form

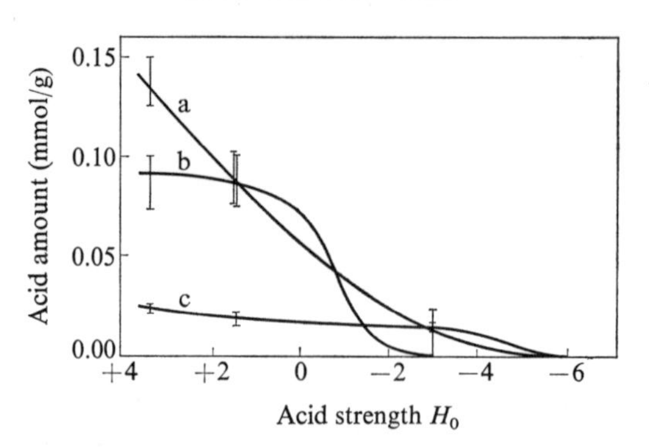

Fig. 4–19 Acid amount *vs.* acid strength for various natural clays
a: attapulgite, b: montmorillonite, c: kaolinite

of linked tetrahedra. They differ from pure silica-alumina in crystal structure, acidity and catalytic activity. Zeolites exist in two forms; "X" with the lower, and "Y" with the higher silica content. A model which has been suggested by Rabo *et al.* for calcium salts of X and Y types is shown in the diagram below.[89]

Ca-X

Ca-Y

The strength and amount of acid on synthesized zeolite Y, together with those of the cation exchanged substance, have been measured by Ukihashi *et al.* using Benesi's amine titration method.[90] As Fig. 4–20 shows, both acid strength and amount are high, whereas zeolite X and the related cation exchanged catalysts have been found to possess a smaller amount of acid, with strengths up to $H_0 = -8.2$. Both the strength and amount of the acid on zeolite X increase after cation exchange (e.g. with Ca^{2+}), but the strength is still less than that of silica-alumina as measured by calorimetric titration with adsorbed ammonia[10] and the *n*-butylamine titration method with H_0 and H_R indicators.[91,92] Table 4–11 gives the acidity distribution of zeolites X as measured by Hirsch-

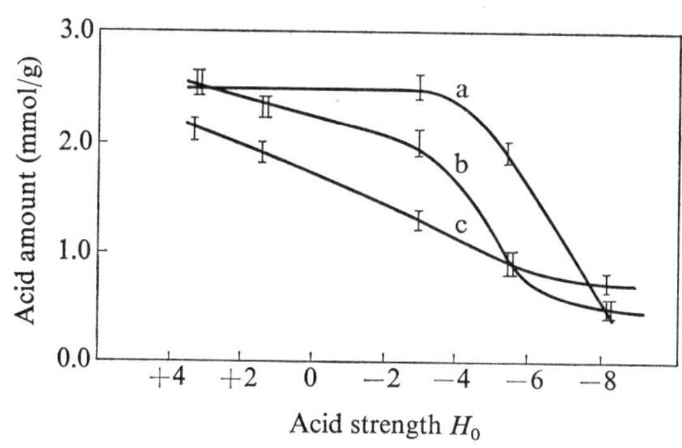

Fig. 4–20 Acid amount *vs.* acid strength for a synthetic Y zeolite, and for two cation exchanged catalysts[90)]
a: the synthetic Y zeolite (H^+)
b: calcium cation exchanged (Ca^{2+})
c: lanthanum cation exchanged (La^{3+})

TABLE 4–11 Acidity distribution of salts of zeolite X

Acid strength (wt% H_2SO_4)	*n*–Butylamine titre, mmol/g stronger than designated H_2SO_4 concentration		
	13X (Na–X)	Ca–X	Li–X
		H_R indicators	
1.2	0	0.50	0.8
36	0	0.29	0.4
50	0	0.17	0.2
68	0	0	0
		H_0 indicators	
3×10^{-4}	0.32	0.35	0.04
48	0	0.26	0.03
72	0	0.03	0

ler.[91)] Both the strength and amount of acid are higher in Ca-X than in Li-X. The total amount of acid (B+L) at $H_0 \leqq +1.5$ on zeolite X subjected to various cation exchanges is listed in Table 4–12, which also gives the catalytic activities for the polymerization of propylene (k_p) and the decomposition of isobutane (k_i).[93)] The correlation between acidity and catalytic activity is not a simple one, but all of the catalytically active substances have acid sites with $H_0 \leqq +1.5$. Richard-

TABLE 4–12 B + L acidity at $H_0 \leqq +1.5$ and activity of cation exchanged zeolite X

Catalysts	Total acidity (mmol/g)	k_p ($g^{-1} \cdot min^{-1}$) $\times 10^3$	k_i ($g^{-1} \cdot min^{-1}$) $\times 10^3$
Na-X	0	0	0
Tl-X	0	0	
H-X	0.23	25	0.32
Mg-X	trace	0	0
Ca-X	0.04	0	0
Sr-X	trace	0	0
Ba-X	0	0	0
Zn-X	0.27	3.3	0.29
Cd-X	trace	0	
Mn-X	present	0	same as Zn–X
Co-X	present	1.5	same as Zn–X
Ni-X	present	6.9	240
La-X	0.39	30	14.5
Ce-X	0.25	33	same as La–X
H-Y	0.56	36	20.2
$SiO_2 \cdot Al_2O_3$ (4:1)	0.5	2.7	0.26
$SiO_2 \cdot Al_2O_3$ (1:4)	0.35	12.5	
Al_2O_3	0.14	0	0.48

son[94] and Eberly[95] suggest that in the case of cation exchanged zeolite Y, the hydrogen in the surface OH groups and in adsorbed water molecules is very readily protonized if the cation has high polarizability, that is if the ratio of the charge of the cation e to its radius r (i.e. e/r, the ionic potential) is large. A tentative acid strength distribution for faujasite has also been derived from the assumption that the acid strength is related to the shift in the stretching-vibration frequency of the OH groups.

The nature of the acid sites on zeolites has recently been investigated by means of infrared techniques (*cf.* 2.3.3). The amount of Brønsted and Lewis acid on magnesium-exchanged hydrogen zeolite Y was obtained from infrared spectra of adsorbed pyridine as shown in Fig. 4–21.[96] The total amount of (B+L) acid as a function of the temperature of heat-treatment is given in Fig. 4–22 curve b. In Fig. 4–22 curve a, the plot of the amount (B+2L) is constant above 600 °C, which suggests that one Lewis acid site is formed by the dehydration of two Brønsted sites. Ward has pointed out from infrared studies on cation

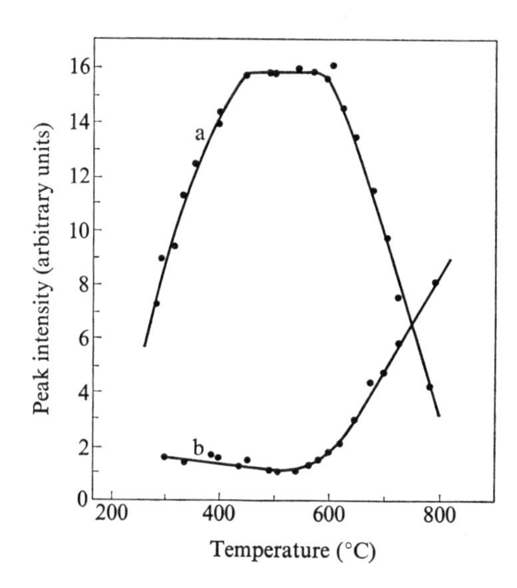

Fig. 4–21 Intensity of infrared absorption bands for pyridine chemisorbed on Brønsted and Lewis acid sites of magnesium hydrogen zeolite Y *vs.* temperature of heat-treatment
a: Brønsted acid sites, b: Lewis acid sites

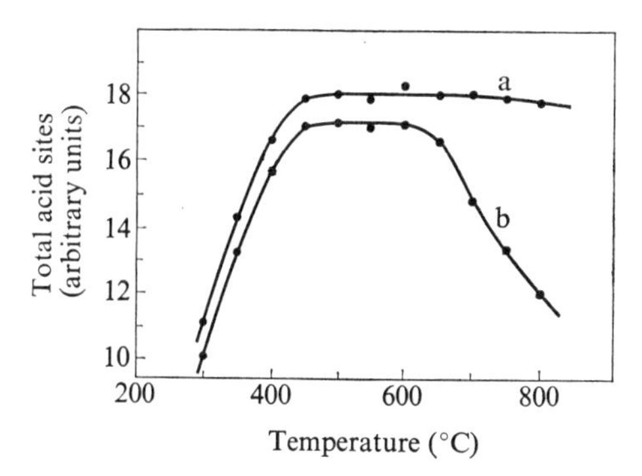

Fig. 4–22 Total acid site population *vs.* temperature of heat-treatment
a: total Brønsted sites plus double total Lewis sites (B+2L),
b: total Brønsted sites plus total Lewis sites (B+L)

exchanged zeolite Y with various cations of the same charge number, that the smaller the ion radius, the stronger the Lewis sites and the larger the amount of Brønsted sites.[96] In addition to the above studies, there have been a number of recent reports of the presence of both Brønsted and Lewis sites on cation exchanged zeolites, and that Brønsted sites are converted into Lewis sites at elevated dehydration temperatures.[95,97–101] Liengme and Hall point out that infrared spectra of pyridine chemisorbed on Lewis acid sites indicate two species of Lewis site; one a weak site caused by a retained sodium ion, and the other a strongly dehydroxylated site.[100] Hattori and Shiba also distinguish between two kinds of Lewis acid, one strong and easily converted to Brønsted type with the addition of a water molecule, and the other weak and highly resistant to conversion.[99] The Lewis sites on NH_4-X, La-X, Ce-X, Ca-X and Sr-X are of the former (strong) type, and those on Zn-X, Mg-X and Mn-X of the latter (weak) type. Acidic OH groups on decationized zeolites (NH_4-X and NH_4-Y) and cation exchanged zeolites are also found to have two infrared bands, at 3,550 and 3,650 cm^{-1}, the OH groups associated with the latter band being the more strongly acidic.[98,100,102]

The structures of acidic and basic sites on decationized zeolites (NH_4-X and NH_4-Y) have been studied by Hall *et al.*[103–5] and Turkevich *et al.*[106–7] The model proposed by Hall *et al.*[103] is shown in the diagram below. On this scheme, an absorption band which was found

```
A      NH4+                                   H+
    O   O   O                             O   O   O
     Al-   Si        ------>   NH3  +      Al-   Si
    O  O  O  O                            O  O  O  O
                                          Brønsted site

B       H+
    O   O   O                 O   HO   O
     Al-   Si       <====>     Al     Si
    O  O  O  O                O  O  O  O

C    O  HO   O                 O   O   O            O         O
  2   Al    Si   --heat-->  H2O +  Al-   Si     +     Al     Si+
     O  O  O  O                O  O  O  O           O  O  O  O
                                Basic site          Lewis site
```

at 3,660 cm^{-1} is attributed to the OH stretching vibration of Si-OH. Addition of ammonia decreases the intensity of this band, and simultaneously that characteristic of NH_4^+ appears. Upon evacuation at

500 °C, the OH band further decreases in intensity and strongly electrophilic sites (Lewis acid sites) can be identified. In diagram B the aluminium atom of three-fold coordination is also considered to be a rather weak Lewis acid site. Turkevich *et al.* have also proposed a similar model for the mechanism by which Brønsted, Lewis, and base sites are formed on decationized zeolite NH_4-Y.[106-7] Their observations of the ESR spectra for adsorbed trinitrobenzene (see Fig. 4–23) show that

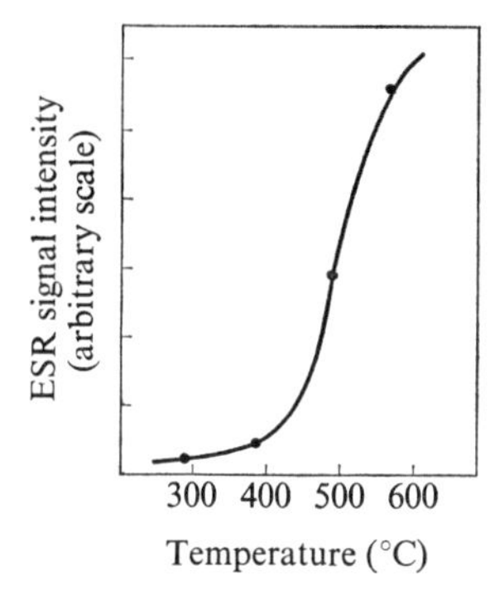

Fig. 4–23 ESR signal intensity for trinitrobenzene adsorbed on zeolite NH_4-Y (giving a measure of the electron donor site population on this decationized sample) *vs.* temperature of evacuation

electron donor sites (Lewis sites in the broadest sense) increase with increasing dehydration temperatures.[107] The diagram below illustrates the mechanism proposed by Ward for the formation of acid sites on zeolite Y, exchanged with a divalent alkaline earth metal.[96] Rabo *et al.*

Na^+ Na^+ — Si, Al^-, Si, Si, Al^- framework $\xrightarrow{M^{2+}}$ $M^{2+}(OH_2)_n$ — Si, Al^-, Si, Si, Al^- framework

$\downarrow$

MO — Si, Al^-, Si, Si^+, Al framework $\xleftarrow[> 500\,°C]{heat}$ $M(OH)^+$, H–O — Si, Al^-, Si, Si, Al framework

suggest that the Lewis acid sites of Le-Y and Ce-Y zeolites are the same as those of decationized zeolite Y with oxygen-deficient sites.[108]

4.4 Metal sulfates and phosphates

Ordinary metal sulfates and phosphates have no intrinsic surface acidity, and become solid acids only after appropriate physical treatment. This is in sharp contrast with the intrinsically acidic hydrogen sulfates. The neutral salt requires activation of some kind before its surface can show significant acidic properties. Any process involving heat, compression, or irradiation which induces some imperfection on the otherwise regular crystal surface may effectively activate the substance. Such metal sulfates and phosphates form a whole new family of solid acids with moderate acid strength, showing characteristic catalytic activity and selectivity. The nature of the acid sites has been investigated for some metal sulfates.

4.4.1 Metal sulfates

Fig. 4–24 shows the amount of acid per unit surface area plotted against the acid strength for samples of nickel sulfate heat-treated at various temperatures.[109] It has been confirmed that the solid acidity is intrinsic, not arising from impurities.[110] It is apparent that acid sites having strengths greater than $H_0=-3$ appear on the surface of nickel sulfate when it is heat-treated at temperatures between 150 and 464 °C. Such sites develop during the course of dehydration, reaching a maximum when the H_2O is in the proportion of something less than 1 mole per mole of $NiSO_4$, and disappearing completely in the anhydrous form. This point will be discussed below in connection with Fig. 4–25 and 4–29. Since the amount of surface acid in a given H_0 range is found from the difference between the two *n*-butylamine titres using the two indicators which bracket that H_0 range, the sharp "step" which develops in the curves at temperatures much above 400 °C in Fig. 4–24 shows a remarkable increase in the number of acid sites at intermediate acid strengths $H_0=+4.0$ to $+3.3$. Such phenomena were not observed in the case of cupric sulfate. Indeed, it is characteristic of the latter that a large number of acid sites having $H_0=+6.8$ to $+4.0$ appears on heating at a relatively low temperature (100 °C).[111]

Fig. 4–25 gives the acid amounts of nickel sulfate at various acid strengths plotted against the temperature of heat-treatment.[109] The amount of acid at all acid strengths initially increases with increasing

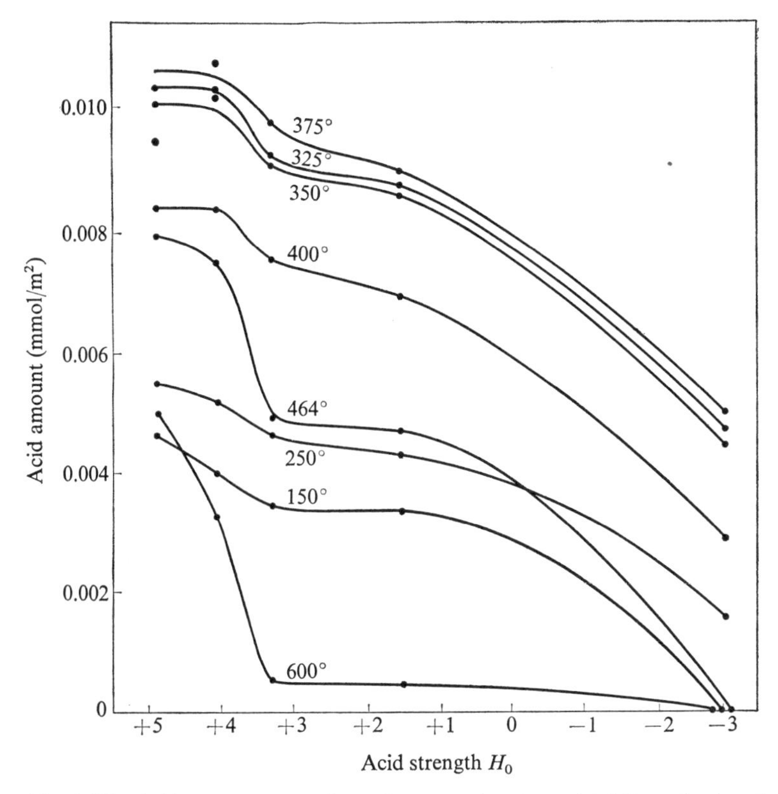

Fig. 4–24 Acid amount per unit surface area (as determined from the butylamine titre) *vs*. acid strength for samples of nickel sulfate heat-treated at various temperatures (°C)

temperature of heat-treatment, reaches a maximum at about 350 °C, and subsequently decreases, as does the catalytic activity (dotted line). Cupric sulfate also shows similar changes in acidity, but the maxima appear at 300 °C.[111] The corresponding maxima for $Al_2(SO_4)_3$, $ZnSO_4$, $MgSO_4$ and $CdSO_4$ are at 400, 180, 260 and 250 °C respectively as shown in Fig. 4–26. Kawaguchi *et al.* found that the maxima obtained by the *n*-butylamine method coincide with those obtained by the DPPH method mentioned in 2.3.2.[112] The acidic properties of other metal sulfates have not been fully investigated, but are essentially similar to those of nickel sulfate. Table 4–13 lists the acidic properties of various metal sulfates.

Since metal sulfates do not change the colours of basic indicators having $pK_a=-5.6$ or -8.2, their acid strengths are relatively low when compared with those of silica-alumina ($pK_a \leqq -8.2$), natural clays

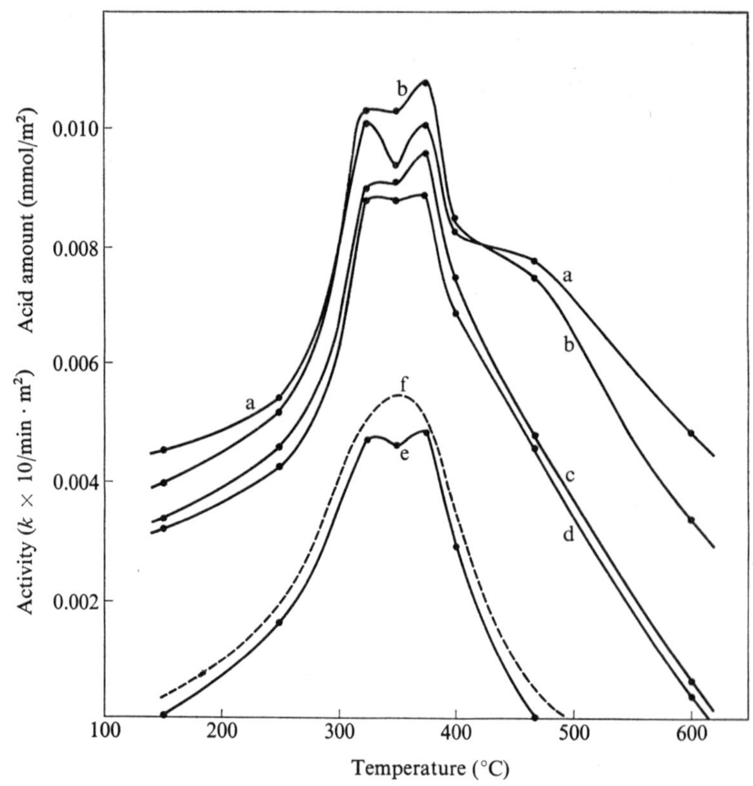

Fig. 4–25 Acid amount and catalytic activity *vs.* temperature of heat-treatment for nickel sulfate at various acid strengths (*cf.* Fig. 4–24)
a: $H_0 \leqq +4.8$, b: $H_0 \leqq +4.0$, c: $H_0 \leqq +3.3$,
d: $H_0 \leqq +1.5$, e: $H_0 \leqq -3.0$, f: (---) catalytic activity

($pK_a = -5.6$ to -8.2) or Lewis acids such as aluminium chloride, boron fluoride, etc., but it is evident from Table 4–13 and Fig. 4–24 and 4–25 that they are not as weak as titanium oxide or zinc sulfide (*cf.* 4.1).

The acidic properties of metal sulfates can be changed by means other than heating, for example by inducing crystal imperfections through compression of the solid. The effects of pressure on the acidic properties of some metal sulfates have been investigated by Ogino and Kawakami.[114,117] They found that compression at 3,000 kg/cm² increases the amount of acid at $H_0 = +3.3$ and $+1.5$ for $CdSO_4 \cdot 8H_2O$, $Ce_2(SO_4)_3 \cdot 8H_2O$, $Fe_2(SO_4)_3 \cdot xH_2O$, $Al_2(SO_4)_3 \cdot 18H_2O$ and $KHSO_4$. The surface acidity of aluminium sulfate at $H_0 = +3.3$ as a function of the applied pressure is given in Fig. 4–27.[114] From this it is evident that the acid sites are generated in proportion to the applied pressure up

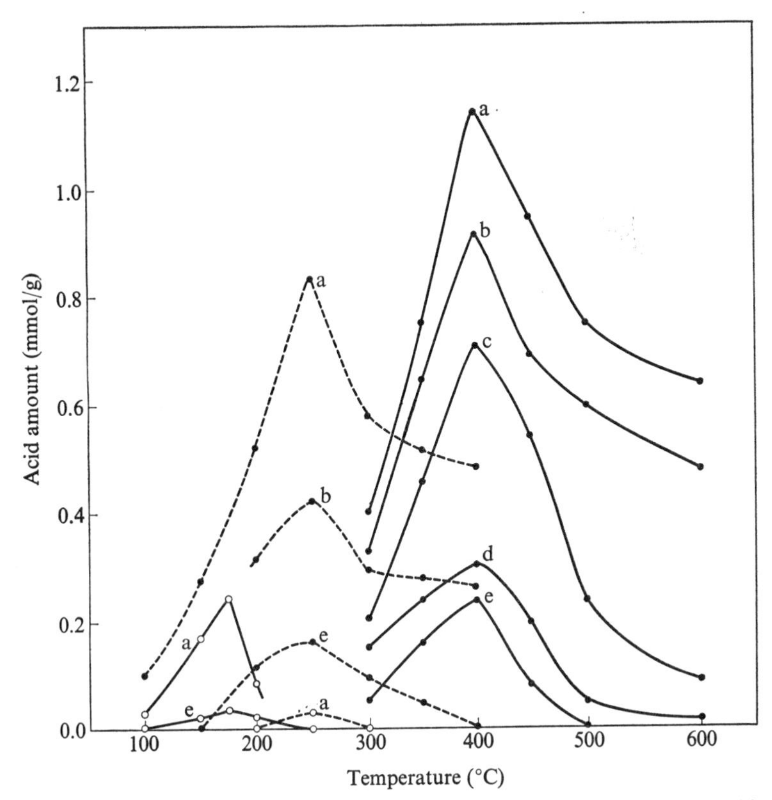

Fig. 4–26 Acid amount *vs*. temperature of heat-treatment for four metallic sulfates at various acid strengths

a: $H_0 \leqq +6.8$, b: $H_0 \leqq +4.8$, c: $H_0 \leqq +3.3$, d: $H_0 \leqq +1.5$, e: $H_0 \leqq +0.8$; (—●—) $Al_2(SO_4)_3$, (--●--) $MgSO_4$, (—○—) $ZnSO_4$, (--○--) $CdSO_4$

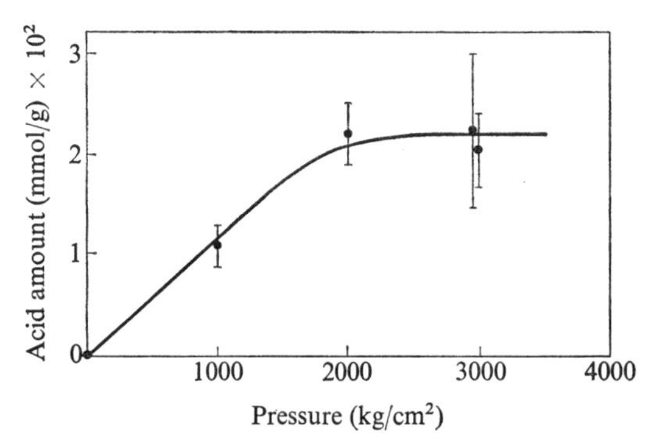

Fig. 4–27 Acid amount (as determined from the *n*-butylamine titre) *vs*. applied pressure for aluminium sulfate at $H_0 \leqq +3.3$

TABLE 4–13 Acidic properties of metal sulfates[49, 109–16)]

Original metal sulfates	Temperature of heat-treatment (°C)	Acidity (mmol/g) at various pK_a							
		+6.8	+4.8	+4.0	+3.3	+1.5	+0.8	−3	−5.6
$CuSO_4 \cdot 5H_2O$	300†	0.227	0.169	0.126	0.126	0.125	—	0.0724	—
$MgSO_4 \cdot 7H_2O$	250†	0.83	0.42	—	—	—	0.16	—	—
$CaSO_4 \cdot 2H_2O$	230†	—	—	0.035	0.018	0	—	—	—
$SrSO_4$	140	—	R	—	R	—	—	—	—
$BaSO_4$	140	—	R	—	R	—	—	—	—
$ZnSO_4 \cdot 7H_2O$	175†	0.25	—	0.16	0.088	0.035	0.04	—	—
$CdSO_4 \cdot 8H_2O$	250†	—	0.03	—	R	R	—	—	—
$Ce_2(SO_4)_3 \cdot 8H_2O$	25	—	—	—	R	R	—	—	—
$Al_2(SO_4)_3 \cdot 18H_2O$	350†	1.1	0.91	—	0.70	0.30	0.24	0	0
$Cr_2(SO_4)_3 \cdot 18H_2O$	100†	—	—	0.73	0.56	0.16	0	—	—
$MnSO_4 \cdot 7H_2O$	230†	—	R	0.035	0.035	0.018	0	—	—
$FeSO_4 \cdot 7H_2O$	200†	—	0.48	0.42	0.38	0.22	0	—	—
$Fe_2(SO_4)_3 \cdot xH_2O$	100†	—	—	2.3	2.2	0.64	0.40	—	0
$CoSO_4 \cdot 7H_2O$	300	—	—	—	0.35	—	—	—	—
$NiSO_4 \cdot 7H_2O$	350†	0.230	0.126	0.126	0.111	0.107	—	0.056	0

† Temperature of heat-treatment giving maximum acidity
R indicates colour change of basic indicator to its conjugate acid form.

to a certain saturation value. $CaSO_4 \cdot 2H_2O$ developed a fairly strong acid strength under compression, and the acidity at $-5.6 < H_0 \leqq -3$ was found to correlate with the extent of lattice distortions.[117]

Hydrated nickel sulfate which has not been heat-treated, and initially shows no acidity, develops definite acidity at $+3.3 \geqq H_0 > -3$ after gamma irradiation (^{60}Co source, total dosage 5.7×10^7 röntgen). Similar irradiation of partially dehydrated $NiSO_4$ heat-treated at 150 ~ 450 °C does increase acidity within the same range of H_0 ($+3.3 \geqq H_0 > -3$), although it leaves the acid amount at $H_0 \leqq -3$ unchanged, and actually reduces the amount at $+4.8 \geqq H_0 > +3.3$ (see Fig. 4–28).[118] The anhydrous form, heat-treated at 550 °C, does not undergo any change in its acidic properties as a result of irradiation. Acid sites formed on nickel sulfate by this means do not decay, although the decay of acid sites on silica gel which had been formed by ionizing radiation has already been described in 4.1.3. A related observation concerns the enhanced catalytic activity of air-dried magnesium sulfate containing radioactive sulfur as compared with the normal non-radioactive catalyst in the dehydration of cyclohexanol.[119]

It is appropriate at this point to discuss the structure of the acid sites on metal sulfates. Fig. 4–25 and 4–26 have shown how remarkably

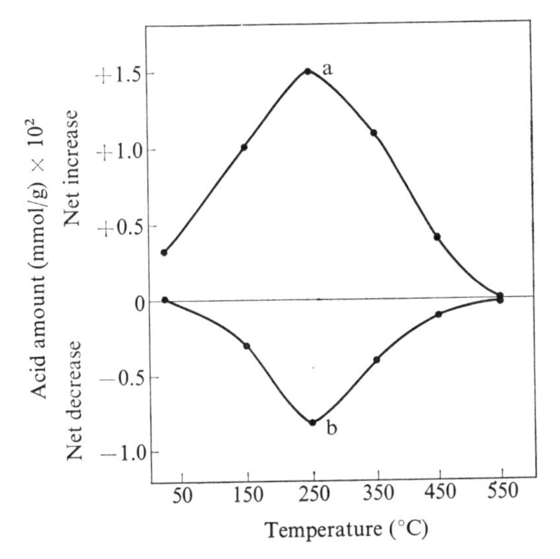

Fig. 4–28 Acid amount *vs.* temperature of heat-treatment for $NiSO_4 \cdot xH_2O$ following γ-irradiation
a: net increase in acid amount at $+3.3 \geqq H_0 > -3$,
b: net decrease in acid amount at $+4.8 \geqq H_0 > +3.3$

the acidic properties of metal sulfates are affected by heating. Nickel sulfate, for example, normally exists as a heptahydrate below 31 °C. Heating will rapidly drive off most of the water, and at 150 °C the monohydrate predominates. Continued heating to about 300 °C causes little further loss of water, but at temperatures much above this the remaining water quickly evaporates, leaving anhydrous nickel sulfate at 400 °C (see Fig. 4–29). Comparison of Fig. 4–25 with Fig. 4–29 indicates that the surface acidity rises with loss of water, and reaches a maximum at about 0.5 mol water per mole of the sulfate. The acidity then drops sharply upon further heating and continues to fall until conversion to the anhydrous structure is complete. Since the amount of water seems to be a critical factor for maximum acidity, the structural features of nickel sulfate have been studied with special reference to the role of water. On the basis of studies with infrared, electron spin resonance, X-ray, nuclear magnetic resonance and Mössbauer effect techniques, Tanabe *et al.* have proposed that the acid centre is formed by an empty orbital of the nickel ion which appears in an incompletely dehydrated metastable transition structure (see (II) in diagram below).[115,120] This configuration is intermediate between the mono-

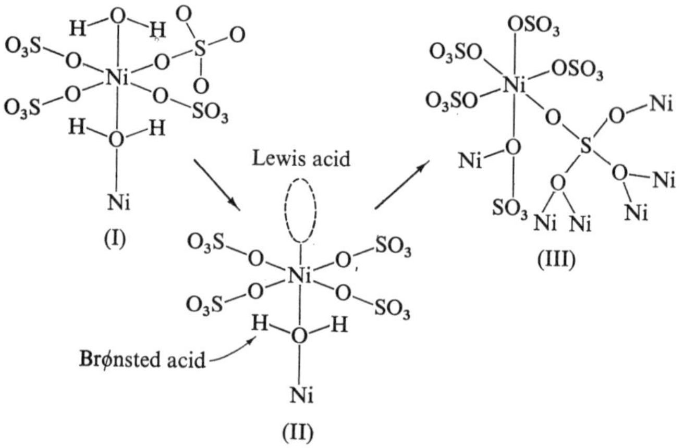

(I)

Lewis acid

Brønsted acid

(II)

(III)

hydrate (I) and anhydrous (III) forms. In this transitional form, nickel is pentavalent, and has a vacant sp^3d^2 orbital. This vacant orbital, and the resultant affinity for an electron pair, accounts for the Lewis acid properties of nickel sulfate and its catalytic activity. The suggested structure is both strained and unstable, although the crystal network and the retained water do have some stabilizing influence. The Brønsted acidity arises from two sources. One is the water which is coordinated directly with a nickel ion in the above transitional form. The nickel

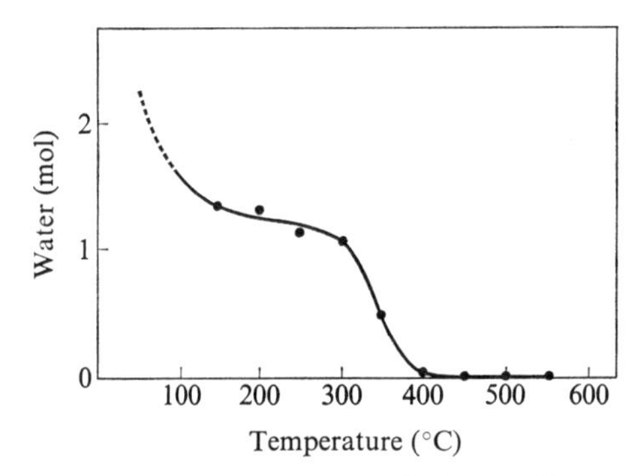

Fig. 4–29 Water content of nickel sulfate *vs.* temperature of heat-treatment

tends to attract the oxygen atom, thus freeing a hydrogen ion. The other source is the surface water, acidified by the inductive effect of the neighbouring cationic Lewis acid centres. The temperature dependence of the Brønsted and Lewis acidity, as derived from the infrared spectra of adsorbed pyridine (*cf.* 2.3.3), is shown in Fig. 4–30.[121] Brønsted acid first appears when the vacant orbital of the nickel ion is formed by dehydration, and the amount increases progressively with increase in dehydration temperature. Since, however, the amount of water of hydration decreases as the temperature of heat-treatment is increased, a temperature is eventually reached at which the Brønsted acidity begins to decline. Lewis acidity also increases with increasing heat-treatment temperature, but only begins to decline when the metastable structure

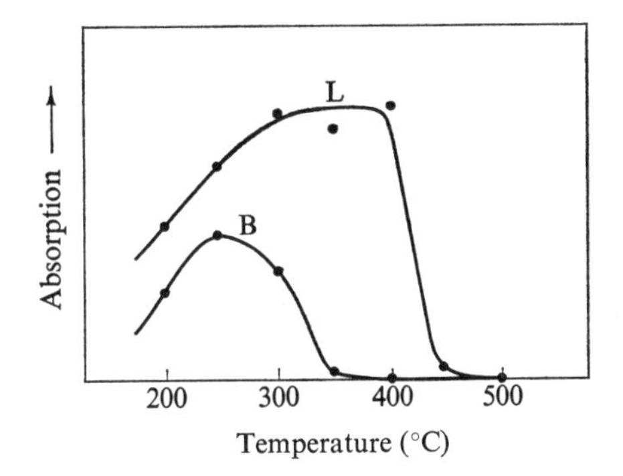

Fig. 4–30 Infrared absorption on Brønsted and Lewis acid sites *vs.* heat-treatment temperature

L: absorption at 1,425 cm^{-1} (Lewis acid),
B: absorption at 1,520 cm^{-1} (Brønsted acid)

with the vacant orbital collapses and changes to the stable anhydrous structure at higher temperatures. The sum of the two acidity curves gives the total acid amount which can be measured by the amine titration method (*cf.* Fig. 4–25). $MgSO_4$, $MnSO_4$, $FeSO_4$, $CoSO_4$, $CuSO_4$ and $ZnSO_4$, which have monohydrate structures similar to that of nickel sulfate, are also considered to have acid centres which are comparable in structure with those of nickel sulfate.[122]

The structure described above is associated with those relatively strong acid sites which are formed on the transitional configuration between the monohydrate and the anhydrate. Ben-Dor and Margalith have studied the structures of the polyhydrates of seven metal sulfates by infrared and differential thermal analysis, correlating the data with those from X-ray diffraction experiments.[123] They explain the relatively weak acid properties of the monohydrates of $MgSO_4$, $CoSO_4$, $NiSO_4$, $ZnSO_4$ and $CdSO_4$ in terms of the special linkage of the water molecule to the central metal ion. This central ion is thought to be octahedrally coordinated, two of the bonds being $M–H_2O$. The hydrogen atoms bonded to oxygen atoms from two different $SO_4{}^{2-}$ groups are readily able to break free as protons. However, for metal sulfates which have little or no prior heat-treatment, the individual acid strengths, though weak, will be largely dependent upon the properties of each cation, in particular its tendency to accept an electron, which is in turn closely related to its electronegativity. Tanaka *et al.* contend that the acid properties of a plain metal sulfate with little heat-treatment depend largely upon the cation electronegativity X_i, where

$$X_i = a + 2bZ = (1 + 2bZ/X_0)X_0 \tag{4.1}$$

and X_0 is the electronegativity of the neutral atom (Z, the charge of the metal ion, is zero) after Pauling. Taking the value of b/X_0 ($= b/a$, both constants) in the equation as unity, a generalized expression for the electronegativity of a metal ion may be derived as follows:[124]

$$X_i = (1 + 2Z)X_0 \tag{4.2}$$

Values calculated from this expression appear to be in good accord with the variations in catalytic activity of several metal sulfate catalysts in the hydration of propylene and the polymerization of aldehyde.

We may note here the similarity between the role of the ionic crystal plane in stimulating acid-catalyzed ionic reactions and that of a polar solvent and an ionic salt in speeding an ionic reaction in solution (by lowering the free energy of activation electrostatically).[125] Pertinent to this conception of a "polar surface" is the recent observation that the basic indicator malachite green-*o*-carboxylic acid lactone, when chemisorbed on $MgSO_4$, $CaSO_4$ and $BaSO_4$ and on LiCl, NaCl and KCl, gives a zwitterion by opening the lactone ring.[126]

$$\overline{C-O-C}=O \xrightarrow{M^+X^-} \overline{C^{\oplus\ominus}O-C}=O$$

In short, the sulfate catalyst has assumed the dual function of an intrinsic acid and a polar medium. This must be responsible, at least in part, for the superior catalytic activity of $NiSO_4$ in many reactions. One other aspect which is often overlooked, but which is receiving increasing attention, is the importance not only of surface acid sites but also of surface base sites in enhancing selectivity. This matter will be discussed in the following chapter.

4.4.2 Metal phosphates

Solid metal phosphates have been used for some time as catalysts in the polymerization of olefins to low polymers, a process which is known to be catalyzed by many other solid acids. It is only recently, however, that surface acidity has been observed in the metal phosphates themselves. The acidic properties of some metal phosphates which have been measured by Tada *et al.* are shown in Fig. 4–31.[127] The strengths

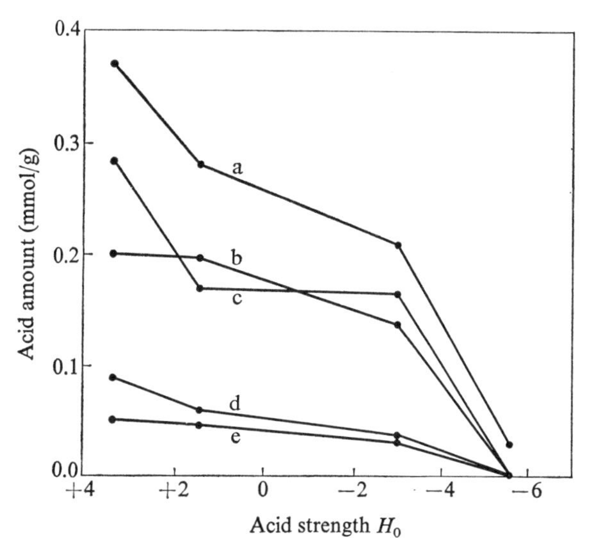

Fig. 4–31 Acid amount *vs.* acid strength for various metal phosphates
a: BPO_4, b: $AlPO_4$, c: $FePO_4$, d: $Cu_3(PO_4)_2$, e: $Ni_3(PO_4)_2$

are rather less than those of $SiO_2 \cdot Al_2O_3$, etc., but not as weak as those of TiO_2, ZnS, etc., being in almost the same range of moderate strengths as the metal sulfates. The experimental observations of Kagiya *et al.* on the effect of calcination temperature on the crystallinity, infrared adsorption, and acid amount at $H_0 \leqq +3.3$, etc. of zirconium phosphate are given in Table 4–14.[128] The infrared band at 1,635 cm^{-1},

TABLE 4–14 Change of acidity etc. of zirconium phosphate with heat-treatment

Temperature of heat-treatment (°C)	Change of weight (%)	Crystal-linity (X-ray)	Infrared absorption		Surface area (m^2/g)	Acidity (mmol/g)
			1,635 cm^{-1}	750 cm^{-1}		
100	100	—	s	—	106	0.68
300	49	—	m	—	78	0.48
500	47	—	w	sh	67	0.23
700	46	—	w	w	37	0.17
900	46	+	—	w	11	0
1,000	46	++	—	m	—	—
1,100	46	+++	—	m	3	—
1,250	46	+++	—	m	3	—

w: weak, m: medium, s: strong, sh: shoulder

which is characteristic of the hydrates, disappears above 900 °C as further reduction in weight becomes negligible, while the band at 750 cm^{-1}, which is assigned to the P-O-P bond, first appears at temperatures above 700 °C and becomes very prominent above 1,000 °C, indicating transformation to zirconium pyrophosphate. A peak which characterizes the pyrophosphate also appears in the X-ray chart at temperatures above 900 °C. Both the observed amount of acid and the polymerization rate of ethylene oxide decrease with rise of calcination temperature, but the molecular weight of the polymer increases remarkably above 900 °C.

4.5 Others

This category includes carbonates, hydroxides, halides, mounted acids and active carbons, the broad features of which are described below.

4.5.1 Carbonates, hydroxides and halides

Metal carbonates such as Na_2CO_3, K_2CO_3, $KHCO_3$, $KNaCO_3$, $(NH_4)_2CO_3$, $BaCO_3$ and $SrCO_3$ are known to change bromothymol blue from yellow to blue.[45,49] Recently, Ogino *et al.* have shown that phenolphthalein is adsorbed on some metal carbonates to give its basic colour, and that the basicity is increased by compression (Table 4–15).[129] This table also gives the basicity of some metal hydroxides.

TABLE 4–15 Effect of applied pressure on the basic properties of carbonates and hydroxides

Carbonates and hydroxides	Applied pressure	
	0 kg/cm²	4,500 kg/cm²
$MgCO_3$	slightly pink	pink
$CaCO_3$	very slightly pink	pink
$SrCO_3$	slightly pink	pink
$BaCO_3$	pink	pink
Na_2CO_3	red violet	intense red violet
$Na_2CO_3 \cdot H_2O$	red violet	intense red violet
K_2CO_3	violet	violet
$KHCO_3$	very slightly pink	pink
$(NH_4)_2CO_3$	colourless	colourless
$Mg(OH)_2$	colourless	slightly pink
$Ca(OH)_2$	red violet	intense red violet
$Zn(OH)_2$	slightly pink	slightly pink
$Ba(OH)_2 \cdot 8H_2O$	slightly violet	slightly violet

Colours are those of adsorbed phenolphthalein

We have already seen how remarkably the basicity of some hydroxides may be increased by heat-treatment (*cf.* 4.1.2).

Many metal halides which have been widely used as strong acid catalysts are thought to show surface acidity. In fact, $AlCl_3$, $SbCl_3$, $SnCl_2 \cdot 2H_2O$, $FeCl_3 \cdot 6H_2O$, $ZnCl_2$, $CaCl_2$, $HgCl_2$, $CrCl_3 \cdot xH_2O$, Cu_2Cl_2 and CaF_2 do adsorb basic indicators such as methyl red and dimethyl yellow to give their red (acidic) colours.[45,49] $PbCl_2$ and $HgCl_2$ both produce the red colour in methyl red, but not in dimethyl yellow. Heat-treatment produces an increase in the amount of acid over that of the untreated substance in the cases of $HgCl_2$ (10 min under vacuum at 180~200 °C), $SnCl_2$ (12 min at 150~170 °C) and $CaCl_2$ (10 min at 180~270 °C). The acidity of $AlCl_3$, however, after an initial increase upon heating under vacuum for 9 min at 180 °C, actually decreases when heated further for 4 min at 280~320 °C.[45] This observation seems to parallel the fact that the catalytic activity of $AlCl_3$ in the polymerization of isobutene is minimal both when it contains water and when it is completely dehydrated, but at a maximum when a small amount of water is present.[130]

The amount and strength of the surface acid for several samples of $TiCl_3$ have been measured by using the amine titration method for coloured samples (2.2.2).[131] Fig. 4–32 gives data for $TiCl_3$ preheated under vacuum at 200 °C, where AR refers to a sample (a Stauffer Chemical product) said to have been reduced by metalic aluminium, and HR to one reduced by hydrogen. ARA and HRA refer respectively

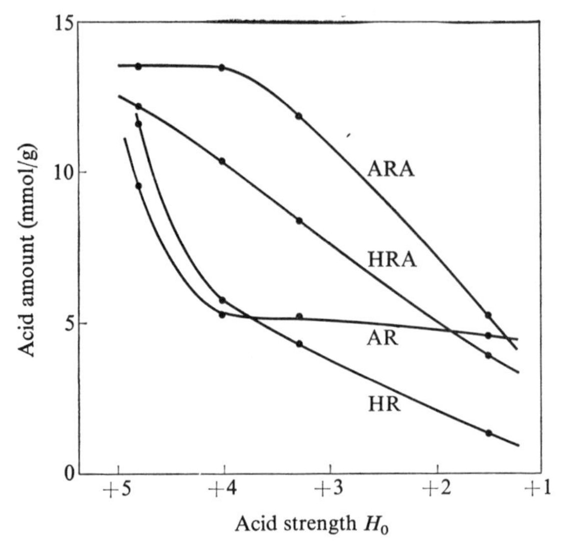

Fig. 4–32 Acid amount (as determined from the *n*-butylamine titre) *vs.* acid strength for $TiCl_3$ heat-treated under vacuum at 200 °C
See text for the significance of ARA, HRA, etc.

to the activated forms. It may readily be seen that the acid amount is anomalously large in comparison with those of the various solid acids mentioned thus far. The amount increases with increasing temperature (25, 200, 300 °C) for each of the four samples. Of the four, the acidic properties of ARA are by far the strongest, which correlates with its pronounced catalytic activity in propylene polymerization to low polymers noted by Keii *et al.*[132] Titanium trichloride-triethyl aluminium (a Ziegler-Natta catalyst) also possesses a large amount of acid at $H_0 \leqq +3.3$, approximating to that of titanium chloride alone at the same acid strength.[131] This observation helps us to understand how it is that the addition of traces of some basic organic compounds can accelerate propylene polymerization to high polymers.[133] It may be justifiable to associate an increase in the catalytic activity of Ziegler-Natta catalysts with coverage of only the stronger acid sites by basic compounds.

Calcium fluoride not only shows acidity as previously listed, but also has basic properties. When phenolphthalein is adsorbed on films of CaF_2 and BaF_2 which have been sublimed under a regimen strictly eliminating all traces of water, they give a bright red (475 mμ) and a red-violet (536 mμ) colour respectively.[134] The larger bathochromic shift of BaF_2 is attributed to a reduction in the effective screening of the negative charge of the outer layer of F^- anions by the positive charge on the underlying Ba^{2+} cations as compared with that due to the Ca^{2+} cations, which are more compact.[134] *p*-Nitrophenol is also adsorbed on CaF_2 and BaF_2, with absorption maxima at 365 and 413 mμ, which are close to but not coincident with that of the anion $^{-}O\text{-}Ph\text{-}NO_2$ in aqueous NaOH (400 mμ). Since the absorption maximum of *p*-nitrophenol in aqueous HCl which occurs at 316 mμ is characteristic of the neutral molecule, the spectrum seems to suffer an unusually large bathochromic displacement. The red shift in the case of BaF_2, however, is even stronger than that for the *p*-nitrophenolate anion. Perhaps we should attribute the large bathochromic shifts observed in these adsorption systems to the stronger adsorption of the excited molecule, with a consequent large reduction of its excitation level. The adsorbed excited *p*-nitrophenol molecule has been represented as a limiting resonance structure interacting with a surface F^- anion, *viz.*,

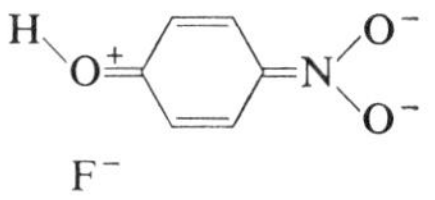

$$\underset{F^-}{H\text{-}\overset{+}{O}}=C_6H_4=N\begin{matrix} O^- \\ O^- \end{matrix}$$

where the ring lies flat on the surface.[135-36] Terenin pointed out the merits of this scheme as a representation of the orientation of the adsorbed molecule in the light of data for the change in electronic distribution of some polar molecules upon excitation to the first single π-π^*, or to the n-π^* levels.[137]

4.5.2 Mounted Acids

Phosphoric acid mounted on diatomaceous earth, silica gel, quartz sand or titanium oxide has been used in the catalysis of the low polymerization and hydration of olefins, the alkylation of certain aromatic compounds, etc., but only recently have the acidic properties of the catalyst surfaces been determined. The acid strength and amount for several acids mounted on silica gel (Davison Grade 70, 60$\sim$200 mesh) measured by Benesi's amine titration method (2.2.1) are shown in Fig. 4–33.[88] The order of acid strength is $H_3BO_3 < H_3PO_4 < H_2SO_4$. The acid amount within various ranges of acid strength derived from Fig. 4–33 are given in Table 4–16. Solid phosphoric acid has a large amount of acid at $-3 > H_0 > -5.6$, whereas solid sulfuric acid is rich in acid sites at $+4 > H_0 > -3$ and $-5.6 > H_0 > -8.2$.

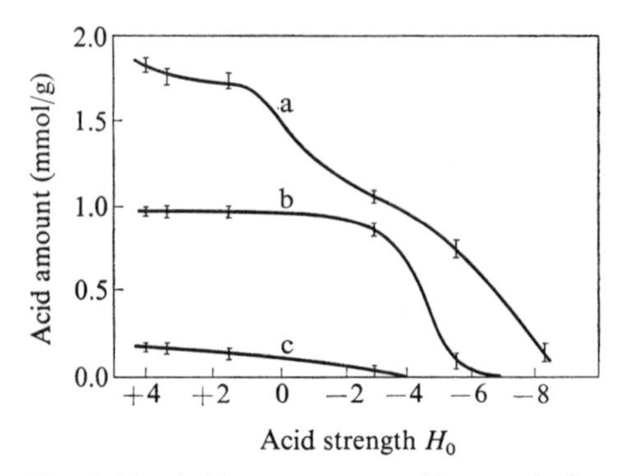

Fig. 4–33 Acid amount *vs.* acid strength for various acids mounted on silica gel dried at 120 °C
a: H_2SO_4, b: H_3PO_4, c: H_3BO_3

TABLE 4–16 Acid amounts of mounted acids

Mounted acids	Surface area (m²/g)	Acid amount at various H_0 ranges (mmol/g)				
		$+4 > H_0 > -3$	$-3 > H_0 > -5.6$	$-5.6 > H_0 > -8.2$	$H_0 < -8.2$	total H_0 range
H_3BO_3	357	0.15	0.02	0	0	0.17
H_3PO_4	330	0.10	0.80	0.07	0	0.97
H_2SO_4	307	0.80	0.30	0.60	0.15	1.86

Mounted acids are generally thought to lose their acidity when heat-treated at high temperature, but Mitsutani and Hamamoto have found that solid phosphoric acid heat-treated at temperatures as high as 700 ~ 1,200 °C still retains fairly strong acidity as shown by Fig. 4–34.[138] Compared with the catalyst heat-treated at less than 600 °C, catalytic activity in the double-bond isomerization of butene is enhanced, as will be mentioned in the next chapter, but the amount of acid at $H_0 \leqq -5.6$ drops to almost zero.

The acid sites of mounted acids are naturally classified as Brønsted type, although there is still some uncertainty over whether this should apply to those heat-treated at high temperatures. On the other hand, the anions of mounted acids can sometimes be thought of as acting as

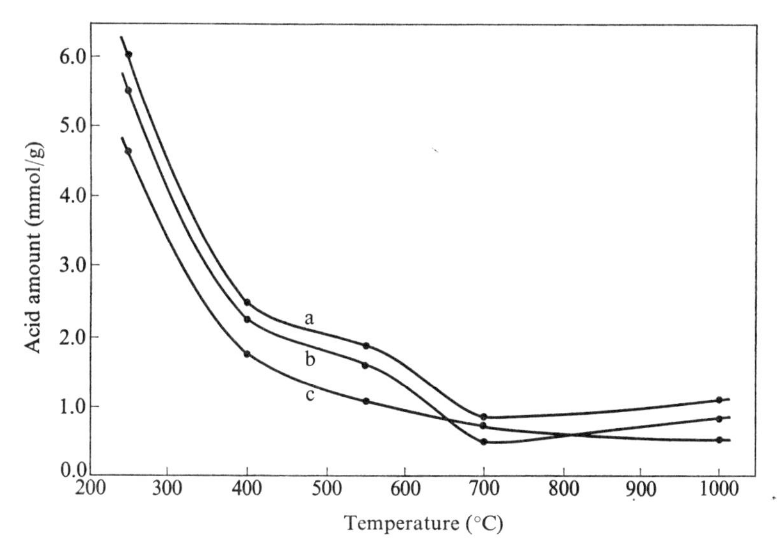

Fig. 4–34 Acid amount *vs.* heat-treatment temperature for solid (i.e. mounted) phosphoric acid at various acid strengths
a: $H_0 \leqq +4.0$, b: $H_0 \leqq +3.3$, c: $H_0 \leqq +1.5$

Brønsted bases, so that the active centres in this kind of catalyst for certain reactions are thought to be a combination of Brønsted acid and base sites. This point, too, will be fully dealt with in the following chapter.

4.5.3 Active carbons

Pure sugar charcoal has both acidic and basic properties. As shown in Fig. 3–5, the amount of base measured by the acid adsorption method (3.2.3) reaches a maximum when the charcoal is heat-treated at about 800~900 °C, whereas the acid amount reaches its maximum at about 400 °C.[139] The amounts of both acid and base are extremely large: the maximum acidity values at 400 °C are 1.99 and 1.75 mmol/g for adsorbed NH_4OH and NaOH respectively, and the maximum basicity values at 810 °C are 48.2, 26.5 and 0.43 mmol/g for adsorbed C_6H_5COOH, CH_3COOH and HCl respectively. The number of basic sites on charcoal activated with nitrous oxide as measured by the exchange method (3.2.2) also attains a maximum when heat-treated at 800 °C, as has already been indicated in Table 3–3.[140] The amounts of base on carbon prepared from analytical filter paper (Toyo Filter Paper No. 2) by baking at 700 °C after immersion in $ZnCl_2$ solution, and by drying after washing in a stream of CO_2 at various temperatures after immersion in NH_4Cl solution, are given in Table 4–17.[141] The maximum amount is once again observed for the sample heat-treated at 800 °C.

TABLE 4–17 Basicity of carbon activated with NH_4Cl

Activation temperature (°C)	300	400	500	600	700	800	900
Surface area (m^2/g)	1.61	1.64	1.62	1.58	1.63	1.62	1.64
Basicity pH	6.7	6.8	6.6	7.2	7.8	9.0	8.5

The basic and acidic properties of activated carbons are considered to be due to the presence of chemisorbed oxygen or an oxygen complex on the surface.[142-3] Shilov attributes the basic properties to the formation of an oxygen complex as shown below.[143]

```
 O   O   O
 ‖   ‖   ‖
 C   C   C
/ \ / \ / \
   C   C
  / \ / \
```

Frumkin regards the charcoal as functioning somewhat in the manner of a reversible electrode, which can be either hydrogen- or hydroxyl-like, dependent upon the temperature of activation.[144] This may be illustrated as follows:

$$\rangle C\langle^{H^+} + OH^- \longleftarrow \rangle C\langle^{H}_{OH} \longrightarrow \rangle C\langle_{OH^-} + H^+$$

High temperature activation, basic solution — Low temperature activation, acidic solution

The appearance of basic properties on nitrous oxide-activated charcoal is considered to be due to one of the three following processes. The solution becomes alkaline 1) due to the release of OH^- by anion exchange as below;

$$C\overline{OH} + Cl^- \longrightarrow C\overline{Cl} + OH^-$$

where $\overline{OH}$ and $\overline{Cl}$ denote the hydroxide and chlorine on the surface of the carbon respectively, or 2) due to a drop in the proton concentration in the solution as protons are selectively attracted to the π-electron of the carbon, or 3) due to the drop in proton concentration in the solution arising from attraction by anions adsorbed on the carbon.[145] Of the above three possibilities 2) should probably be excluded, for as Table 4–18 shows,[140] the basicity varies with the kinds of anion

TABLE 4–18 Basicity of active carbon

Temperature of heat-treatment (°C)	Basicity pH		
	KCl	KBr	KI
untreated	7.1	7.3	7.9
800	9.9	10.1	10.9

which are used for its measurement. Thus it appears that the basicity is due either to anion exchange or anion adsorption.

REFERENCES

1. E. Echigoya, *Nippon Kagaku Zasshi*, **76**, 1144 (1955).
2. T. Yamaguchi, H. Shimizu and K. Tanabe, unpublished data.
3. M. Ito, K. Matsuura, Y. Yoshii, K. Igarashi and A. Suzuki, *Dai 22-nenkai Kōenyokō-shū* (Japanese), Chem. Soc. of Japan, No. 06420 (1969).
4. A. N. Webb, *Ind. Eng. Chem.*, **49**, 261 (1957).
5. Y. Trambouze, *Compt. Rend.*, **233**, 648 (1951); Y. Trambouze, L. de Mourgues and M. Perrin, *J. Chim. Phys.*, **51**, 723 (1954); *Compt. Rend.*, **236**, 1023 (1953).
6. A. E. Hirschler, *J. Catalysis*, **6**, 1 (1966).
7. H. Pines and W. O. Haag, *J. Am. Chem. Soc.*, **82**, 2471 (1960).
8. E. Echigoya, *Nippon Kagaku Zasshi*, **76**, 1049 (1955).
9. E. P. Parry, *J. Catalysis*, **2**, 371 (1963).
10. H. R. Gerberich, F. E. Lutinski and W. K. Hall, *ibid.*, **6**, 209 (1966).
11. V. C. F. Holm and A. Clark, *Ind. Eng. Chem., Prod. Res. Develop.*, **2**, 38 (1963).
12. T. V. Antipina, O. V. Bulgakov and A. V. Uvarov, *Intern. Congr. Catalysis, 4th, Moscow, Preprints of Papers*, No. 77 (1968).
13. G. -M. Schwab and H. Kral, *Proc. Intern. Congr. Catalysis, 3rd, Amsterdam*, I, No. 20 (1964).
14. See the discussion by Dr. Hair in ref. 13.
15. M. Yamadaya, K. Shimomura, T. Konoshita and H. Uchida, *Shokubai (Tokyo)*, **7**, No. 3, 313 (1965).
16. H. Pines and J. Manassen, *Advances in Catalysis*, vol. 16, p. 49, Academic Press, 1966.
17. S. G. Hindin and S. W. Weller, *J. Phys. Chem.*, **60**, 1501 (1956).
18. J. B. Peri, *ibid.*, **69**, 220 (1965).
19. J. B. Peri, *ibid.*, **69**, 211 (1965).
20. J. B. Peri, *ibid.*, **69**, 231 (1965).
21. E. Echigoya, a dissertation presented to Tokyo Inst. of Technology in partial

fulfillment of the requirements for the Ph.D. degree (1957).
22. E. B. Cornelius, T. H. Milliken, G. A. Mills and A. G. Oblad, *J. Phys. Chem.*, **59,** 809 (1955).
23. O. V. Krylov and E. A. Fokina, *Probl. Kinetiki i Kataliza, Acad. Nauk, USSR*, vol. 8, p. 248, 1955.
24. T. Iizuka and K. Tanabe, unpublished data.
25. J. Take, N. Kikuchi and Y. Yoneda, *Shokubai* (*Tokyo*), **10,** (23rd Symp. Catalysis, Preprints of Papers), 127 (1968).
26. T. Kawaguchi, S. Hasegawa, S. Morikawa and H. Suzuki, *Shōwa 43-nendo Shokubai Kenkyū Happyōkai Kōenyōshi-shū* (Japanese), Catalysis Soc. of Japan, No. 3 (1968).
27. S. Morikawa and T. Kawaguchi, *Dai 22-nenkai Kōenyokō-shū* (Japanese), Chem. Soc. of Japan, No. 06426 (1969).
28. O. V. Krylov, Z. A. Markova, I. I. Tretiakov and E. A. Fokina, *Kinetics and Catalysis* (*USSR*) (*Eng. Transl.*), **6,** 128 (1965).
29. S. Malinowski, S. Szczepanska and J. Sloczynski, *J. Catalysis*, **7,** 67 (1964).
30. G. Kortüm, *Angew. Chem.*, **70,** 651 (1958).
31. H. Zeitlin, R. Frei and M. McCarter, *J. Catalysis*, **4,** 77 (1965).
32. H. E. Zaugg and A. D. Schaffer, *J. Am. Chem. Soc.*, **87,** 1857 (1965).
33. V. Y. Lodin, V. E. Kholmogorov and A. N. Terenin, *Dokl. Akad. Nauk, USSR*, **160,** 1347 (1965).
34. Y. D. Pimenov, V. E. Kholmogorov and A. N. Terenin, *ibid.*, **163,** 935 (1965).
35. R. L. Nelson, A. J. Tench and B. J. Harmsworth, *Trans. Faraday Soc.*, **63,** 1427 (1967); A. J. Tench and R. L. Nelson, *ibid.*, **63,** 2254 (1967).
36. T. Iizuka, H. Hattori, K. Tanabe, Y. Ohno and J. Soma, *Shokubai* (*Tokyo*), **11,** No. 4, 90P (1969).
37. J. E. Mapes and R. R. Eischens, *J. Phys. Chem.*, **58,** 809 (1954); W. A. Pliskin and R. R. Eischens, *ibid.*, **59,** 1156 (1955).
38. L. M. Roev, V. N. Filimonov and A. N. Terenin, *Optika i Spektroskopiya*, **4,** 328 (1958).
39. I. D. Chapman and M. L. Hair, *J. Catalysis*, **2,** 145 (1963).
40. K. Tarama, S. Teranishi, S. Yoshida, H. Honda and S. Taniguchi, *Dai 18-nenkai Kōenyokō-shū* (Japanese), Chem. Soc. of Japan, No. 2409 (1965).
41. C. Barter and C. D. Wagner, *J. Phys. Chem.*, **68,** 2381 (1964).
42. K. Tanabe and T. Yamaguchi, *J. Res. Inst. Catalysis, Hokkaido Univ.*, **11,** 179 (1964).
43. E. Morinari and G. Parravano, *J. Am. Chem. Soc.*, **75,** 5233 (1953).
44. K. Tanabe, C. Ishiya, I. Matsuzaki, I. Ichikawa and H. Hattori, *Dai 23-nenkai Kōenyokō-shū* (Japanese) (Ann. Meeting Chem. Soc. Japan, 23rd, Tokyo, Preprints of Papers), No. 03408 (1970).
45. K. Tanabe and M. Katayama, *J. Res. Inst. Catalysis, Hokkaido Univ.*, **7,** 106 (1959).
46. T. Kawaguchi, S. Hasegawa, K. Kaseda and S. Kurita, *Denshi Shashin* (Japanese), **8,** 92 (1968).
47. S. E. Voltz, A. E. Hirschler and A. Smith, *J. Phys. Chem.*, **64,** 1594 (1960).
48. N. Pernicone, G. Liberti and L. Ersini, *Intern. Congr. Catalysis, 4th, Moscow, Preprints of Papers*, No. 21 (1968).
49. K. Nishimura, *Nippon Kagaku Zasshi*, **81,** 1680 (1960).
50. K. Tanabe and T. Yamaguchi, *J. Res. Inst. Catalysis, Hokkaido Univ.*, **14,** 93 (1966).

51. T. Shiba, M. Sato, H. Hattori and K. Yoshida, *Shokubai* (*Tokyo*), **6,** No. 2, 80 (1964).
52. M. R. Basila, T. R. Kantner and K. H. Rhee, *J. Phys. Chem.*, **68,** 3197 (1964).
53. M. R. Basila and T. R. Kantner, *ibid.*, **70,** 1681 (1966).
54. Y. Trambouze, *J. Chem. Phys.*, **51,** 723 (1954).
55. M. R. Basila and T. R. Kantner, *J. Phys. Chem.*, **71,** 467 (1967).
56. W. K. Hall, H. P. Leftin, F. J. Cheselske and D. E. O'Reilly, *J. Catalysis*, **2,** 506 (1963).
57. T. Shirasaki, M. Okada and F. Suganuma, *Shokubai* (*Tokyo*), **6,** No. 4, 265 (1964).
58. E. Echigoya and H. Niiyama, *ibid.*, **6,** No. 4, 292 (1964).
59. C. J. Plank, D. J. Sibbert and R. B. Smith, *Ind. Eng. Chem.*, **49,** 742 (1957).
60. USSR. Pat. 148,031 (1963).
61. G. M. Panchenkov, *Zh. Fiz. Khim.*, **36,** 1113 (1962).
62. R. J. Mikovsky and P. B. Weisz, *J. Catalysis*, **1,** 345 (1962).
63. A. Mitsutani, T. Eguchi and Y. Hamamoto, *Kogyo Kagaku Zasshi*, **68,** 695 (1965).
64. R. C. Hansford, *Advances in Catalysis*, vol. 4, p. 17, Academic Press, 1952.
65. S. Yamaguchi and S. Tsutsumi, *Nippon Kagaku Zasshi*, **69,** 6 (1948).
66. C. L. Thomas, *Ind. Eng. Chem.*, **41,** 2564 (1949).
67. M. W. Tamele, *Discussions Faraday Soc.*, **8,** 270 (1950).
68. T. H. Milliken, Jr., G. H. Mills and A. G. Oblad, *ibid.*, **8,** 279 (1950).
69. J. D. Danforth, *J. Phys. Chem.*, **59,** 564 (1955).
70. C. J. Plank, *J. Colloid Sci.*, **2,** 413 (1947).
71. B. S. Greensfelder, H. H. Voge and G. M. Good, *Ind. Eng. Chem.*, **41,** 2573 (1949).
72. A. G. Oblad, S. G. Hindin and G. A. Mills, *J. Am. Chem. Soc.*, **75,** 4096 (1953).
73. R. G. Haldeman and P. H. Emmett, *ibid.*, **78,** 2917 (1956).
74. G. A. Mills and S. G. Hindin, *ibid.*, **72,** 5549 (1950).
75. J. K. Lee and S. W. Weller, *Anal. Chem.*, **30,** 1057 (1958).
76. L. Leonard, S. Suzuki, J. J. Fripiat and C. De Kimpe, *J. Phys. Chem.*, **68,** 2608 (1964).
77. W. K. Hall, F. E. Lutinski and H. R. Gerberich, *J. Catalysis*, **3,** 512 (1964).
78. J. B. Peri, *Proc. Intern. Congr. Catalysis, 3rd, Amsterdam*, I, No. 72 (1964).
79. T. Shirasaki, M. Okada, T. Mizutori, S. Hayakawa and S. Hata, *Nippon Kagaku Zasshi*, **85,** 722 (1964).
80. Y. Izumi and T. Shiba, *Bull. Chem. Soc. Japan*, **37,** 1797 (1964).
81. V. A. Dzisko, *Proc. Intern. Congr. Catalysis, 3rd, Amsterdam*, I, No. 19 (1964).
82. H. Bremer and K. H. Steinberg, *Intern. Congr. Catalysis, 4th, Moscow, Preprints of Papers*, No. 76 (1968).
83. V. A. Dzisko, M. S. Borisova, L. G. Karakchiev, A. D. Makarov, N. S. Kotsarenko, R. J. Zusman and L. A. Khripin, *Kinetika i Kataliz*, **6,** 1033 (1965).
84. H. Niiyama and E. Echigoya, *Shokubai* (*Tokyo*), **10,** (23rd Symp. Catalysis, Preprints of Papers), 129 (1968).
85. N. S. Kotsarenko, L. G. Karakchiev and V. A. Dzisko, *Kinetika i Kataliz*, **9,** 158 (1968).
86. S. E. Voltz and S. W. Weller, *J. Am. Chem. Soc.*, **76,** 4701 (1954).
87. Brit. Pat. 718,723 (1954).
88. H. A. Benesi, *J. Phys. Chem.*, **61,** 970 (1957).
89. J. A. Rabo, P. E. Pickert, D. N. Stamires and J. E. Boyle, *Actes Congr. Intern.*

Catalyse, 2ᵉ, Paris, II, No. 104 (1960, Pub. 1961).
90. H. Ukihashi, S. Otoma and Y. Arai, *Dai 20-nenkai Kōenyokō-shū* (Japanese) (Ann. Meeting Chem. Soc. Japan, 20th, Tokyo, Preprints of Papers), No. 11209 (1967).
91. A. E. Hirschler, *J. Catalysis*, **2,** 428 (1963).
92. C. J. Norton, *Ind. Eng. Chem. Process Design Develop.*, **3,** 230 (1964).
93. T. Shiba, *Japan–U.S.A. Seminar Catalytic Sci., Tokyo and Kyoto, Preprints of Papers from Japan*, No. J-6-1 (1968); T. Nishizawa, H. Hattori, T. Uematsu and T. Shiba, *Intern. Congr. Catalysis, 4th, Moscow, Preprints of Papers*, No. 55 (1968).
94. J. T. Richardson, *J. Catalysis*, **9,** 182 (1967).
95. P. E. Eberly, Jr., *J. Phys. Chem.*, **72,** 1042 (1968).
96. J. W. Ward, *J. Catalysis*, **9,** 225 (1967); **10,** 34 (1968); **11,** 238, 251 (1968).
97. B. V. Romanovsky, K. S. Tkhoang, K. V. Topchieva and L. I. Piguzova, *Kinetika i Kataliz*, **7,** 841 (1966).
98. T. R. Hughes and H. M. White, *J. Phys. Chem.*, **71,** 2192 (1967).
99. H. Hattori and T. Shiba, *J. Catalysis*, **12,** 111 (1968).
100. B. V. Liengme and W. K. Hall, *Trans. Faraday Soc.*, **62,** 3229 (1966).
101. H. A. Benesi, *J. Catalysis*, **8,** 368 (1967).
102. C. L. Angell and P. C. Schaffer, *J. Phys. Chem.*, **69,** 3463 (1965).
103. J. B. Uytterhoven, L. G. Christner and W. K. Hall, *ibid.*, **69,** 2117 (1965).
104. L. G. Larson and W. K. Hall, *ibid.*, **69,** 3080 (1965).
105. R. P. Porter and W. K. Hall, *J. Catalysis*, **5,** 366 (1966).
106. D. N. Stamires and J. Turkevich, *J. Am. Chem. Soc.*, **86,** 749 (1964).
107. J. Turkevich, *Japan–U.S.A. Seminar Catalytic Sci., Tokyo and Kyoto, Preprints of Papers from U.S.A.*, No. A-3 (1968).
108. J. A. Rabo, C. L. Angell and V. Schomaker, *Intern. Congr. Catalysis, 4th, Moscow, Preprints of Papers*, No. 54 (1968).
109. K. Tanabe and R. Ohnishi, *J. Res. Inst. Catalysis, Hokkaido Univ.*, **10,** 229 (1962).
110. K. Tanabe and M. Katayama, *ibid.*, **7,** 106 (1959); *Shokubai (Sapporo)*, **14,** 1 (1957).
111. K. Tanabe and C. Mugiya, *J. Res. Inst. Catalysis, Hokkaido Univ.*, **14,** 101 (1966).
112. T. Kawaguchi, S. Hasegawa, F. Yamamoto and M. Kurita, *Shokubai (Tokyo)*, **10,** (23rd Symp. Catalysis, Preprints of Papers), 140 (1968).
113. K. Tarama, S. Teranishi, K. Hattori and T. Ishibashi, *ibid.*, **4,** No. 1, 69 (1962).
114. Y. Ogino and T. Kawakami, *Bull. Chem. Soc. Japan*, **38,** 972 (1965).
115. I. Toyoshima, K. Tanabe, T. Yoshioka, J. Koizuka and H. Ikoma, *Dai 22-nenkai Kōenyokō-shū* (Japanese) (Ann. Meeting Chem. Soc. Japan, 22nd, Tokyo, Preprints of Papers), No. 06405 (1969).
116. H. Takida and K. Noro, *Kobunshi Kagaku*, **21,** 23, 109 (1964).
117. T. Kawakami and Y. Ogino, *Shokubai (Tokyo)*, **10,** (23rd Symp. Catalysis, Preprints of Papers), 136 (1968).
118. K. Tanabe, T. Iizuka and M. Ogasawara, *J. Res. Inst. Catalysis, Hokkaido Univ.*, **16,** 567 (1968).
119. C. B. Amphlett, *Chem. Ind. (London)*, 249 (1965).
120. T. Takeshita, R. Ohnishi, T. Matsui and K. Tanabe, *J. Phys. Chem.*, **69,** 4077 (1965).
121. H. Hattori, S. Miyashita and K. Tanabe, *Dai 22-nenkai Kōenyokō-shū* (Japa-

nese) (Ann. Meeting Chem. Soc. Japan, 22nd, Tokyo, Preprints of Papers), No. 06404 (1969).

122. J. Coing-Boyat and G. Bassi, *Compt. Rend.*, **256**, 1482 (1963).
123. L. Ben-Dor and R. Margalith, *Inorganica Chim. Acta*, **1**: **1**, 49 (1967).
124. K. Tanaka, A. Ozaki and K. Tamaru, *Shokubai* (*Tokyo*), **6**, No. 4, 262 (1964); K. Tanaka and A. Ozaki, *J. Catalysis*, **8**, 1 (1967); *Bull. Chem. Soc. Japan*, **40**, 1723 (1967).
125. G.-M. Schwab and K. Garves, *Chem. Ber.*, **98**, 1608 (1965).
126. G. Kortüm and J. Vogel, *ibid.*, **93**, 706 (1960).
127. A. Tada, Y. Yamamoto, M. Ito and A. Suzuki, *Dai 22-nenkai Kōenyokō-shū* (Japanese) (Ann. Meeting Chem. Soc. Japan, 22nd, Tokyo, Preprints of Papers), No. 06421 (1969).
128. T. Kagiya, M. Sano, T. Shimidzu and K. Fukui, *Kogyo Kagaku Zasshi*, **66**, 1893 (1963).
129. W. Osanami, S. Baba, T. Kawakami and Y. Ogino, *Dai 22-nenkai Kōenyokō-shū* (Japanese) (Ann. Meeting Chem. Soc. Japan, 22nd, Tokyo, Preprints of Papers), No. 06424 (1969).
130. A. G. Evans and M. Polanyi, *J. Chem. Soc.*, **1947**, 252; P. H. Plesch, M. Polanyi and H. A. Skinner, *ibid.*, **1947**, 257.
131. K. Tanabe and Y. Watanabe, *J. Res. Inst. Catalysis, Hokkaido Univ.*, **11**, 65 (1963).
132. T. Keii, T. Takagi and S. Kanetaka, *Shokubai* (*Tokyo*), **3**, No. 2, 210 (1961).
133. M. Nakano, S. Matsumura and H. Sasaki, *Dai 9-kai Kōbunshi Tōronkai Kōenyōshi-shū* (Japanese), No. 12A5, Soc. Polymer Sci., Japan (1960).
134. J. H. de Boer, *Z. Physik. Chem.*, **B16**, 397 (1932); **B17**, 161 (1932); J. H. de Boer and J. F. H. Custers, *ibid.*, **B25**, 225, 238 (1934); J. F. H. Custers and J. H. de Boer, *Physica*, **1**, 265 (1934); **3**, 407 (1936).
135. J. H. de Boer and G. M. M. Houben, *Koninkl. Ned. Akad. Wetenschap., Proc., Ser.* **B54**, 421 (1951).
136. J. H. de Boer, *Advances in Colloid Science*, vol. 3, p. 1, 1950.
137. A. N. Terenin, *Advances in Catalysis*, vol. 15, p. 227, Academic Press, 1964.
138. A. Mitsutani and Y. Hamamoto, *Kogyo Kagaku Zasshi*, **67**, 1231 (1964).
139. A. King, *J. Chem. Soc.*, **1937**, 1489.
140. E. Naruko, *Kogyo Kagaku Zasshi*, **67**, 2019 (1964).
141. E. Naruko, *Bull. Univ. Osaka Prefect., Ser.* **A13**, 119 (1964).
142. N. Shilov, H. Schatunowskaja and K. Tschmutov, *Z. Physik. Chem.*, **A149**, 211 (1930); **A150**, 31 (1930); J. H. Wilson and T. R. Bolam, *J. Colloid Sci.*, **5**, 550 (1950); B. R. Puri, D. D. Singh, J. Nath and L. R. Sharma, *Ind. Eng. Chem.*, **50**, 1071 (1958); A. King, *J. Chem. Soc.*, **1934**, 22; S. Weller and T. F. Yound, *J. Am. Chem. Soc.*, **70**, 4155 (1948).
143. N. Shilov, *Z. Physik. Chem.*, **A148**, 233 (1930).
144. A. Frumkin, *Kolloid-Z.*, **51**, 123 (1930).
145. A. Shlygin, A. Frumkin and V. Medvedovskii, *Acta Physicochim.*, **4**, 911 (1936).

Chapter 5

Correlation between Acid-Base Properties and Catalytic Activity and Selectivity

It has been shown in the preceding chapter that a surprisingly large number of solids have surface acidic and/or basic properties. These solid acids and bases have been usefully employed as catalysts in various acid-base catalyzed reactions. The present chapter surveys the correlations that exist between the solid acid-base properties and their catalytic activities and selectivities. The mechanisms of heterogeneous reactions and the characteristics of the catalytic action of solid acid-base catalysts are also discussed and compared with those of homogeneous catalysts by means of a combination of kinetic and structural studies.

5.1 Solid acid catalysis

In this section we first examine the correlation of the catalytic activity of various solid acids with the amount, strength, and nature (Brønsted or Lewis type) of their acid sites, and then discuss the characteristics of solid acid catalysis, particularly the selectivity of these catalysts.

5.1.1 Correlation between acid amount and catalytic activity

Good correlations have been found in many cases between the total amount of acid (Brønsted plus Lewis type, usually measured by the amine titration method) and the catalytic activities of solid acids. For example, the rates of both the catalytic decomposition of cumene[1] and the polymerization of propylene[1,2] over $SiO_2 \cdot Al_2O_3$ catalysts were found to increase with increasing acid amounts at strength $H_0 \leqq +3.3$. Fig. 5–1 gives results for propylene polymerization with a series of $SiO_2 \cdot Al_2O_3$ samples at 200 °C. The total amount of acid for members of the series as measured by amine titration using dimethyl yellow ($pK_a = +3.3$) increases at first with increasing alumina content. The acid amount reaches a maximum somewhere between alumina contents of 10.3% (F in Fig. 5–1) and 25.1% (G), the value for G being rather lower than that for F. Thus, the fact that the polymerization

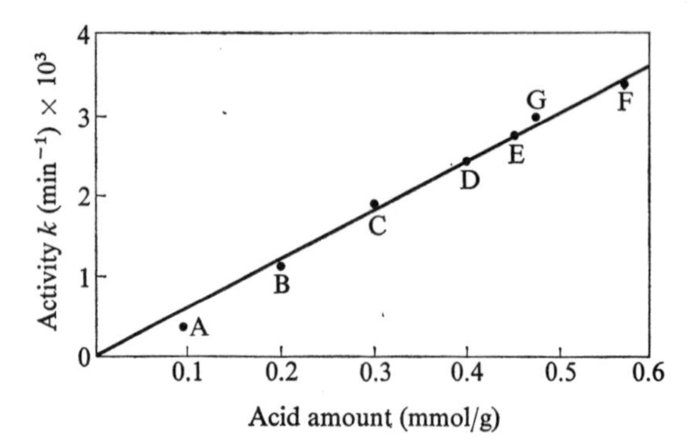

Fig. 5–1 Propylene polymerization activity *vs.* acid amount for a series of $SiO_2 \cdot Al_2O_3$ catalysts

Catalyst	A	B	C	D	E	F	G
Al_2O_3 (wt%)	0.12	0.32	1.04	2.05	3.56	10.3	25.1

activity increases in the same order as the acid amount, i.e. A, B, C, D, E, G and F as shown in the figure, indicates that the total amount of (B+L) acid is a critical factor governing polymerization activity. The situation is a little more complex in the case of cumene cracking with $SiO_2 \cdot Al_2O_3$ catalysts at 500 °C, see Fig. 5–2: the correlation, though evident, is non-linear.[1] Approximately linear relationships have, however, been demonstrated between the acid amount for $SiO_2 \cdot Al_2O_3$ catalysts as measured by the quinoline adsorption method and their cracking activity for volatile petroleum fractions.[3–7] A relationship has also been found to hold between acidity and activation

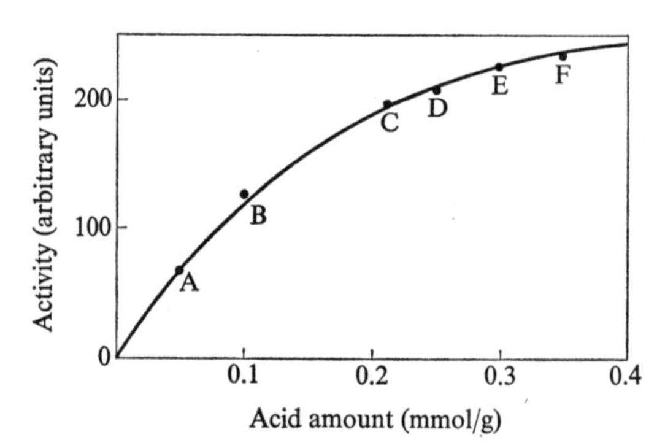

Fig. 5–2 Cumene decomposition activity at 500 °C *vs.* acid amount for a series of $SiO_2 \cdot Al_2O_3$ catalysts

Catalyst	A	B	C	D	E	F
Al_2O_3 (wt%)	0.12	0.32	1.04	2.05	3.56	10.3

energy. In Fig. 5–3, the activation energy for formic acid decomposition is plotted against the acidity of $SiO_2 \cdot Al_2O_3$ determined by amine titration using *p*-dimethylaminoazobenzene.[8] The curve confirms that there is a definite relationship between the activation energy and the acidity. Catalysts A to H were prepared by the vapour phase hydrolysis of a homogeneous mixture of aluminium chloride and silicon tetrachloride, the alumina contents being as given. It is significant that pure silica falls neatly into place on the curve. The anomalous values for the commercial catalyst I show that there must be some other factor in addition to acidity which plays a role in this reaction, for although its acidity is undoubtedly high (0.478 mmol/g), its activation energy is still somewhat higher than might be expected.

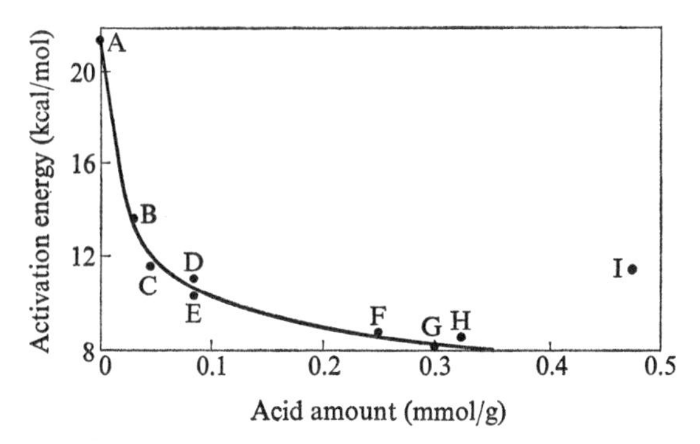

Fig. 5–3 Formic acid decomposition activation energy *vs.* acid amount for a series of $SiO_2 \cdot Al_2O_3$ catalysts

Catalyst	A	B	C	D	E	F	G	H	I
Al_2O_3 (wt% dry basis)	100	0.0	44.9	36.6	33.3	6.7	11.0	5.5	13.0 (commercial sample)

Uchida and Imai have found that both the ethylene polymerization activity of $SiO_2 \cdot Al_2O_3$ and its acid amount are increased by the addition of nickel oxide.[9] Morikawa, Shirasaki *et al.* have shown that the catalytic activity and selectivity of $SiO_2 \cdot Al_2O_3$ catalysts can be enhanced by the adsorption of various cations.[10,11] In fact, the catalytic activity of the nickel ion exchanged catalyst is far greater than that of the unexchanged catalyst in the disproportionation of toluene, the level of activity depending to a very large extent upon the kind and amount of the exchanged cations.[11]

We have seen in Chapter 4 how remarkably the acidic properties of alumina itself change with the degree of dehydration. Alumina's catalytic activities in a range of reactions such as the isomerization of hydrocarbons, the polymerization of olefins, etc., have all been attributed to the acidic properties of the surface.[12–19] Similar relationships have been found between the acidity of Al_2O_3 as measured by the adsorption of pyridine, ammonia, or water vapour, and the polymerization activity for ethylene[17] and propylene.[18]

The relationship between the acidity (B+L) at $H_0 \leqq +1.5$ of zeolite X exchanged with various cations and the corresponding activity in the polymerization of propylene and the decomposition of isobutane has already been given in Table 4–12. Each of the catalysts that displays catalytic activity does in fact possess acid sites of strength $H_0 \leqq +1.5$.[20] Correlations have also been established between the acidity at $H_0 \leqq +3.3$ of zeolite X exchanged with divalent cations such as Cd, Mg, Ca, Sr, and Ba, and the catalytic activity in the decomposition of diisobutylene at 300 °C and the polymerization of isobutylene at 350 °C, by the studies of Matsumoto and Morita.[21] As shown in Fig. 5–4, the relationship between the amount of acid and the conversion rate of diisobutylene into $C_1 \sim C_4$ products is essentially linear. In the case of isobutylene polymerization, although the activity does increase with acidity, the relationship is not linear (see Fig. 5–5). In the former case, the amount of quinoline required to deactivate the catalysts completely was found to be rather larger than the amount of acid at $H_0 \leqq +3.3$ as determined by *n*-butylamine titration.[21] This implies that the weaker acid sites at $H_0 > +3.3$ are also catalytically active for the decomposition of diisobutylene. The pronounced catalytic activity displayed by zeolites in the cracking of cumene has also been shown to be dependent upon the large number of

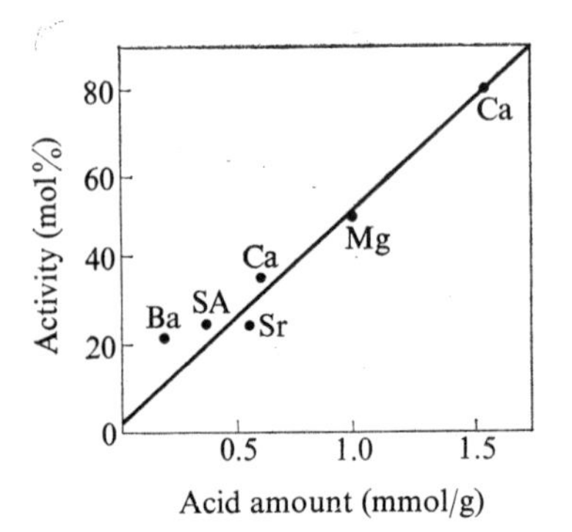

Fig. 5–4 Diisobutylene conversion activity at 300 °C *vs.* acid amount at $H_0 \leqq +3.3$ for cation exchanged zeolites X Silica-alumina (SA above) is included for comparison

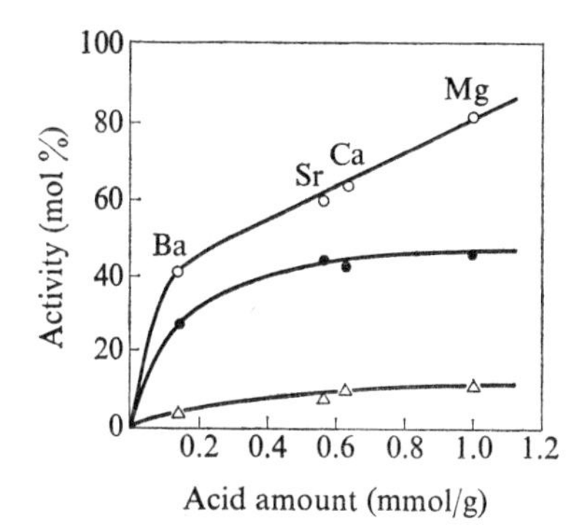

Fig. 5–5 Isobutylene conversion activity at 350 °C *vs.* acid amount at $H_0 \leqq +3.3$ for cation exchanged zeolites X
(○) conversion of isobutylene,
(●) yield of liquid product,
(△) yield of trimer and higher polymers

acid sites, but not specifically upon their acid strength.[22]

The catalytic activity of $Al_2O_3 \cdot B_2O_3$ in the transalkylation of aromatic compounds is found to be higher than that of $SiO_2 \cdot Al_2O_3$.[23] This higher activity seems to be due essentially to the higher concentrations of acid sites on $Al_2O_3 \cdot B_2O_3$ (*cf.* Fig. 4–15). It is also found that $Al_2O_3 \cdot B_2O_3$ shows high activity for the Beckmann rearrangement of cyclohexanone oxime.[24] The activity of $SiO_2 \cdot ZrO_2$ in the dehydration of alcohols and the cracking of cumene is directly proportional to the amount of acid on the surface.[25] Fig. 5–6 shows the linear relationship between catalytic activity in the decomposition of ethyl and isopropyl alcohols and the acidity at $H_0 \leqq -8.2$ (again by amine titration). The points for samples of different chemical composition and calcined at different temperatures all lie on the same straight line. This indicates that the decrease in the activity of $SiO_2 \cdot ZrO_2$ catalysts after heat-treatment is due to a decrease in the number of acid sites rather than to a change in their nature. A similar linear dependence of activity on acid amount is observed in the case of the dehydration of isopropyl alcohol on $SiO_2 \cdot MgO$ catalysts of varying composition.[25] The close correlation between the surface acidity of $MoO_3 \cdot Fe_2(MoO_4)_3$ catalysts at $H_0 \leqq +4.0$ and their activity in the oxidation of methanol to formaldehyde has already been mentioned (*cf.* Fig. 4–17).[26] Tani and Ogino, on the other hand, have found that the activity of $ZnO \cdot Cr_2O_3 \cdot SiO_2 \cdot Al_2O_3$ catalysts for the synthesis of methanol from carbon dioxide and hydrogen largely depends upon the amount of acid within a limited range of fairly

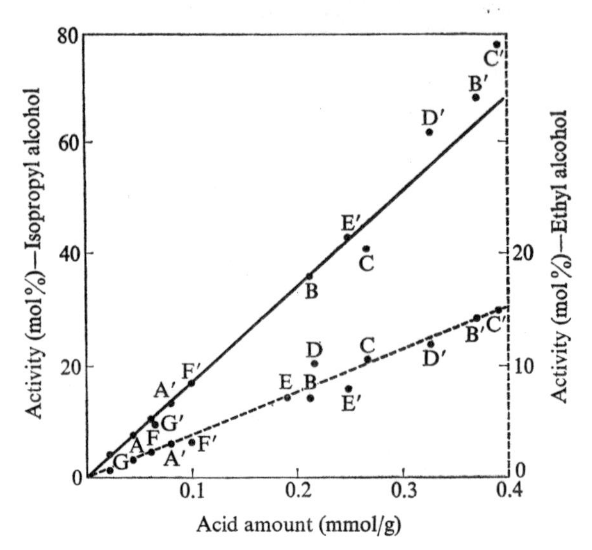

Fig. 5–6 Ethyl and isopropyl alcohol decomposition activity *vs.* acid amount at $H_0 \leqq -8.2$ for a series of $SiO_2 \cdot ZrO_2$ catalysts
(—●—) decomposition of isopropyl alcohol
(--●--) decomposition of ethyl alcohol
A; 850 °C, A′; 500 °C: 10% ZrO_2
B; 800 °C, B′; 500 °C: 45% ZrO_2
C; 800 °C, C′; 500 °C: 50% ZrO_2
D; 800 °C, D′; 500 °C: 55% ZrO_2
E; 750 °C, E′; 500 °C: 71% ZrO_2
F; 500 °C, F′; 435 °C: 85% ZrO_2
G; 500 °C, G′; 425 °C: 92.5% ZrO_2

low acid strengths.[27] Fig. 5–7, the plot of the ratio $\Delta C/\Delta A$ (where ΔC is the incremental change in activity for an increment ΔA in the amount of acid) against acid strength H_0, shows clearly that an increase of the amount of acid within the range $+3.3 \geqq H_0 \geqq +1.5$ will be most effective in enhancing catalytic activity.

Certain reactions catalyzed by metal sulfates and phosphates have revealed good correlations between the acidity measured by amine titration and the catalytic activity. Fig. 4–25, for instance, shows the excellent correlation between the acidity of nickel sulfates and their catalytic activity for the depolymerization of paraldehyde.[28] Close inspection of the figure reveals that nickel sulfate heat-treated at temperatures above 450 °C has a number of sites with acid strengths weaker than $H_0 = +1.5$, but shows little or no catalytic activity. The best correlation for catalytic activity in the depolymerization reaction is with the

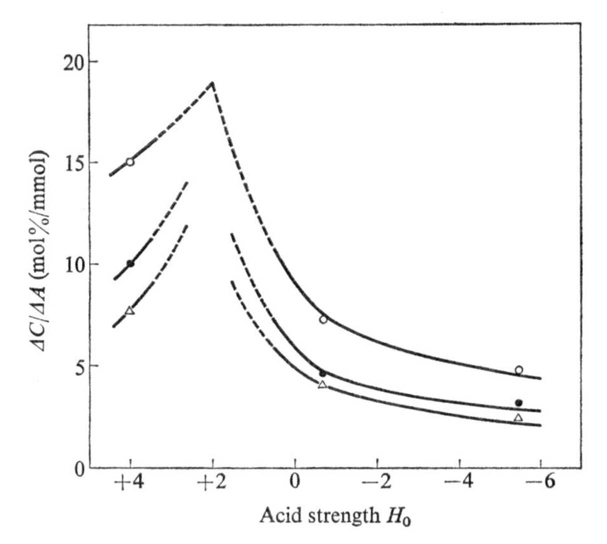

Fig. 5–7 The rate of change of activity with respect to acidity ($\Delta C/\Delta A$) *vs.* acid strength for $ZnO \cdot Cr_2O_3 \cdot SiO_2 \cdot Al_2O_3$ in methanol synthesis
(○) synthesis at 340 °C, (●) at 325 °C, (△) at 305 °C

distribution of acid amount at strengths greater than $H_0 = -3.0$. The fact that nickel sulfate completely loses its catalytic activity when the acid sites at $H_0 \leqq -3.0$ are poisoned with dicinnamalacetone, a basic indicator with $pK_a = -3.0$, is further confirmation of the crucial role of these stronger sites.[29] The reason for the ineffectiveness of acid sites with weaker acid strength ($H_0 > -3.0$) in the reaction is probably to be found in the fact that no paraldehyde molecule will give a conjugate acid at any of these weaker sites, since the pK_a of paraldehyde is about -3.3.[30]

The catalytic depolymerization of paraldehyde by nickel and cupric sulfates proceeds according to enzyme kinetics. In Fig. 5–8, the reciprocal of the depolymerization velocity v (the initial velocity per mmol of acid at $H_0 \leqq -3.0$) is plotted against the reciprocal of the initial concentration [S] of the paraldehyde.[31] The figure shows that the plots lie on two straight lines, indicating that the kinetic data follows the well-known Michaelis-Menten equation:

$$\frac{1}{v} = \frac{K_m}{V_{\max}} \cdot \frac{1}{[S]} + \frac{1}{V_{\max}} \tag{5.1}$$

which is derived from the following scheme for an enzymatic reaction:

$$\text{substrate} + \text{enzyme} \underset{k_2}{\overset{k_1}{\rightleftharpoons}} (\text{substrate-enzyme complex}) \xrightarrow{k_3} \text{product} + \text{enzyme} \quad (5.2)$$

where $K_m = (k_2 + k_3)/k_1$, and $V_{max} = k_3$ [enzyme].

Thus, the depolymerization of paraldehyde with metal sulfates may be represented by a scheme analogous to an enzymatic reaction, as shown below:

$$\text{paraldehyde} + \text{sulfate} \underset{k_2}{\overset{k_1}{\rightleftharpoons}} (\text{paraldehyde-sulfate complex}) \xrightarrow{k_3} \text{acetaldehyde} + \text{sulfate} \quad (5.3)$$

Michaelis' constant is found to be 0.12 and 0.045 respectively for cupric and nickel sulfate.

This mechanism is entirely different from that of the homogeneous reaction catalyzed by a Brønsted acid in a non-aqueous solution. The mechanism for the latter, proposed by Bell *et al.*, is given below for comparison.[32] Note particularly the second order dependence on the acid concentration:

$$P + HA \underset{k_2}{\overset{k_1}{\rightleftharpoons}} PH^+ \cdot A^- \quad (5.4)$$

$$PH^+ \cdot A^- + HA \xrightarrow{k_3} PH_2{}^{2+} \cdot A_2{}^{2-} \xrightarrow{k_4} 3CH_3CHO + 2HA \quad (5.5)$$

where P and HA denote paraldehyde and Brønsted acid respectively, and $k_1 \ll k_2$, $k_3 \ll k_4$, with $k_2 \gg k_3$. The first stage is then in equilibrium, with an equilibrium constant $K = k_1/k_2$, and with the velocity of the overall reaction given by

$$v = k_3 [PH^+ \cdot A^-] [HA] = k_3 K [P] [HA]^2 \quad (5.6)$$

The difference between the rate equations for heterogeneous and homogeneous reactions is clearly evident from Eq. 5.1 and 5.6. The difference between the activation energies is given in Table 5–1.[29,33] In the

TABLE 5–1 Activation energy (E_a) for the depolymerization of paraldehyde

Catalysts	$NiSO_4$	$Al_2(SO_4)_3$	$CuSO_4$	CCl_3COOH
E_a (kcal/mol)	26.9	25.6	21.7	14.3

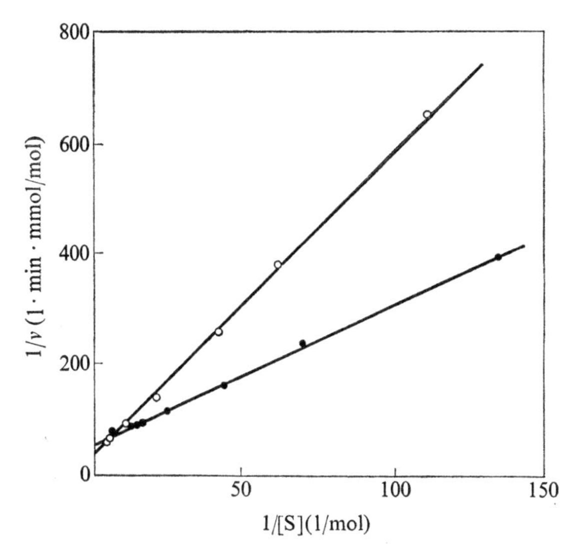

Fig. 5–8 The catalytic depolymerization of paraldehyde by nickel and cupric sulfates, following enzyme kinetics (●) nickel sulfate, (○) cupric sulfate

homogeneous reaction, the transfer of a proton to one or more of the oxygen atoms in the paraldehyde molecule is thought to weaken the C–O bonds. Since, however, it has been shown in the case of the heterogeneous reaction that the sulfate behaves in a similar way to an enzyme, it might be considered that C–O bonds are weakened not only by the adsorption of oxygen atoms of paraldehyde on active centres (both Brønsted and Lewis acid sites) on the catalyst, but also by some interactions between sites other than active centres and some part of the paraldehyde molecule. The first order rate constants (k) of the depolymerization in benzene at 30 °C with several solid acid catalysts and a homogeneous acid are given in Table 5–2, together with their surface

TABLE 5–2 Comparison of the activity of solid acid catalysts with that of a homogeneous acid catalyst

Catalysts	$SiO_2 \cdot Al_2O_3$	$NiSO_4$	$Al_2(SO_4)_3$	$CuSO_4$	$MnSO_4$	CCl_3COOH
Acidity (mmol/g)	0.50	0.09	0.53	0.19	0.11	24 mmol/l
$k \times 10^3$ (min^{-1}. g^{-1})	0.10	6.2	35	2.2	0.05	—
$k \times 10^3$ (min^{-1}. $mmol^{-1}$)	0.20	68	65	12	0.45	0.10

k: rate constant of depolymerization of paraldehyde

acidities at $H_0 \leqq +3.3$.[34] The activity per unit acidity of the solid acids is higher than that of the trichloroacetic acid. Taking the activity per unit acidity (i.e. per mmol of acid) at $H_0 \leqq -3$, which is the range of effective acid strengths for the reaction, the ratio of the activities becomes

$$NiSO_4 : CuSO_4 : SiO_2 \cdot Al_2O_3 = 1{,}100 : 320 : 1$$

Usually the high polymerization of aldehydes is catalyzed not only by bases but also by both Brønsted and Lewis acids. Takida and Noro have found that some metal sulfates do in fact catalyze this polymerization.[35] Measurements of the polymerization activity and acidic properties of solid sulfates of Fe (III), Cr, Zn, Ni, Mn, Mg, Cu, Fe (II) and Ca which had been heat-treated at various temperatures revealed that those catalysts possessing acid sites with $H_0 \leqq +3.3$ are effective in the polymerization. Actually, as shown in Fig. 5–9, there is a fairly close

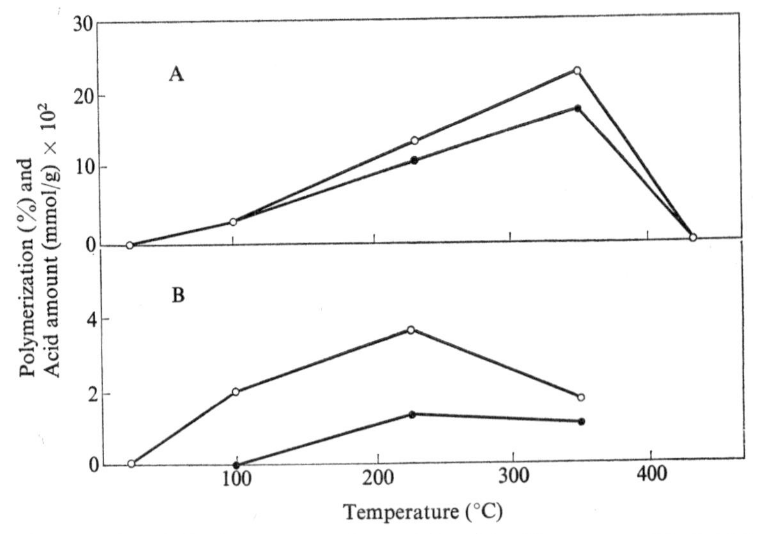

Fig. 5–9 Acetaldehyde polymerization activity and acid amount at $H_0 \leqq +3.3$ for metal sulfate catalysts (A: $NiSO_4$ B: $MgSO_4$) *vs.* temperature of heat-treatment
(○) polymerization activity, (●) acid amount

parallel between the catalytic activity and the acid amount at these strengths. On the other hand, the relationship between the number of activated molecules (given by the polymerization rate W divided by the mean degree of polymerization $\bar{p}$) and the acid amount A is expressed by the equation:

$$\frac{W}{\rho}=\frac{1}{1.27}\cdot A^{1.96}$$

Thus, the number of activated molecules (i.e. the catalytic activity) is approximately proportional to the square of the acidity.

The low polymerization of propylene in the vapour phase was found by Tarama *et al.* to be catalyzed by sulfates of Ni, Co, Fe, Cu, Mn and Zn, the catalytic activity increasing with increase in the number of acid sites having strength $H_0 \leqq +3.3$.[36] Since only the one indicator ($pK_a = +3.3$) was used, however, the range of acid strengths for which the sites are active is not clear. By using a series of indicators, Watanabe and Tanabe have shown that the polymerization rate at 100 °C is proportional to the amount of acid on nickel sulfates at $H_0 \leqq +1.5$ (see Fig. 5–10).[37] This did not hold for acid amounts at $H_0 > +3.3$, $+3.3 < H_0$

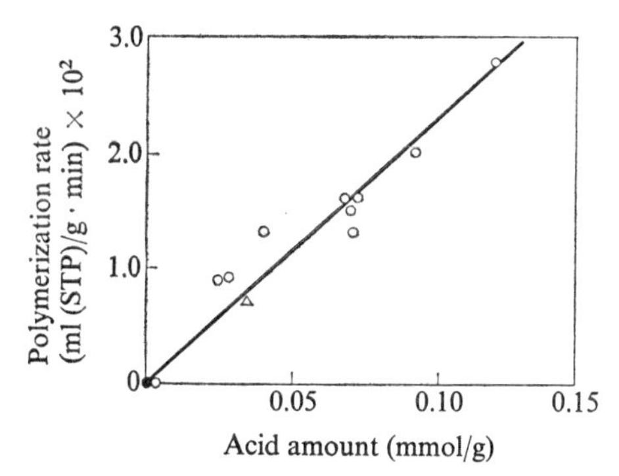

Fig. 5–10 Propylene polymerization rate at 100 °C *vs.* acid amount at $H_0 \leqq +1.5$ on various catalyst samples
(○) $NiSO_4$ heat-treated at various temperatures, (△) $CuSO_4$ heat-treated at 300 °C, (●) TiO_2

$\leqq +4.8$, $+4.0 < H_0 \leqq +4.8$, $-3 < H_0 \leqq +1.5$, and $H_0 \leqq -3$. It is therefore probable that the sites which are catalytically active are those at acid strengths $H_0 \leqq +1.5$. The observation by Tarama, previously quoted,[36] is not incompatible with the latter observation, since there are no acid sites with strengths $+1.5 < H_0 \leqq +3.3$. It was shown in Fig. 5–1 that the propylene polymerization activity of silica-alumina correlates with the amount of acid using *p*-dimethylaminoazobenzene of $pK_a = +3.3$ as the indicator.[2] Since silica-alumina has little or no acid with strength weaker than $H_0 = -8.2$,[38] Johnson's observation is also compatible with Watanabe and Tanabe's results. However, it seems impossible to determine the actual strength of the sites which are active in the polymerization, since the silica-alumina possesses only strong acid sites at $H_0 \leqq -8.2$.

Non heat-treated titanium oxide is known to be a solid acid possessing only weak acid sites, e.g. 0.01 mmol/g of acid at $H_0 \leqq +4.8$, but zero at $H_0 \leqq +1.5$. The rate of polymerization with this solid acid was found to be zero (see Fig. 5–10). This is in accord with the conclusion based on the results with nickel sulfate, that the acid sites at $H_0 \leqq +1.5$ are those catalytically active in polymerization. Aluminium sulfate having $0.7 \sim 0.8$ mmol/g of acid at $H_0 \leqq +1.5$ showed high catalytic activity, but lost this entirely when all the acid sites were poisoned with 0.8 mmol of ammonia per gram of catalyst. This also accords with the above conclusion.[37]

The disproportionation reaction of fluorohalohydrocarbons such as $2CCl_3F \longrightarrow CCl_4 + CCl_2F_2$ is known to be catalyzed by metal salts such as $AlCl_3$, $AlBr_3$, AlF_3, LiF/C and $FeCl_3$/C.[39] Okazaki has shown that the order of their activities for the disproportionation of CCl_3F at 200 °C is $AlF_3 > FeCl_3$, $CaCl_2 > CrCl_3$, $PbCl_2 > CaF_2$, $HgCl_2$, each halide catalyst with the exception of $HgCl_2$ being classified as a hard acid.[40] It was also found that the activity of $NiSO_4$ for the disproportionation reaction correlates with the acidity at $H_0 \leqq +3.3$ as shown in Fig. 5–11. The fact that the reaction is catalyzed by solid acids of metal

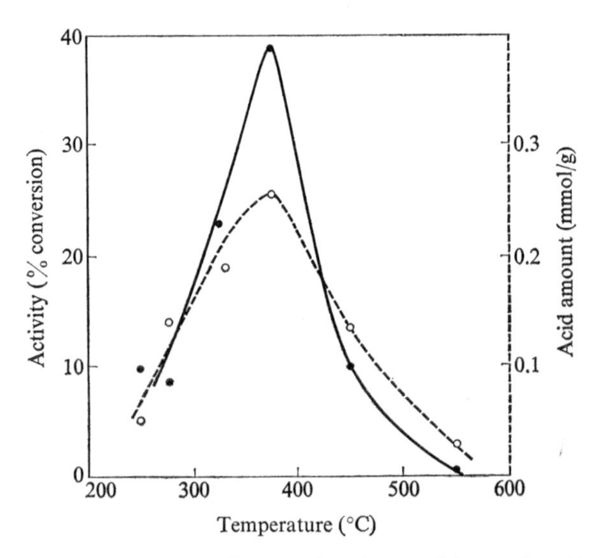

Fig. 5–11 CCl_3F disproportionation activity and acid amount at $H_0 \leqq +3.3$ for $NiSO_4$ *vs.* temperature of heat-treatment
(—●—) activity, (--○--) acid amount

salts MX_n and that it follows Rideal's mechanism led Okazaki to propose the following mechanism for the reaction:[40]

$$CCl_3F + MX_n \longrightarrow CCl_3^{\oplus}[FMX_n]^{\ominus}$$

$$CCl_3^{\oplus}[FMX_n]^{\ominus} + CCl_3F \longrightarrow CCl_4 + CCl_2F_2 + MX_n$$

Correlations between activity and acid amount over limited ranges of acid strength were evident in both the hydration of propylene[41] and the polymerization of isobutylvinylether[42] catalyzed by metal sulfates. The catalytic activities are proportional to the acidities at $-3 < H_0 \leqq +1.5$ and at $-5.6 < H_0 \leqq -3$ respectively (see Fig. 5–12).

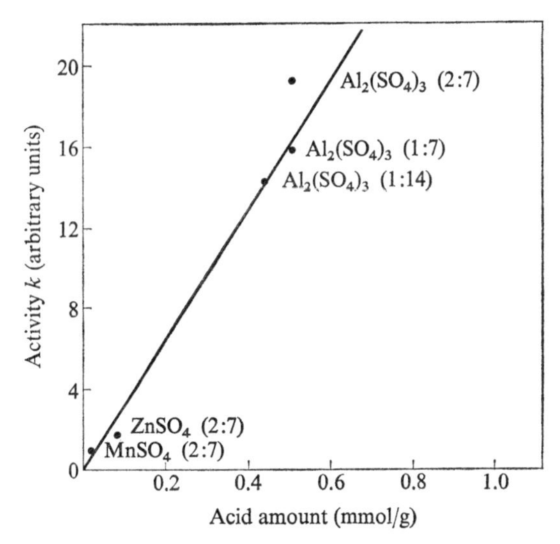

Fig. 5–12 Propylene hydration activity *vs.* acid amount at $-3 < H_0 \leqq +1.5$ for various metal sulfates
Ratios give the proportion of sulfate to silica gel

Various metal phosphates have been found effective as solid acid catalysts for the high polymerization of ethylene oxide,[43] the cracking of volatile fractions of petroleum[44] and the dehydration of isopropyl alcohol.[45] Fig. 5–13 and 5–14 show good correlations between the amount of acid within a definite range of acid strengths and the catalytic activity for phosphates.

It was mentioned in 4.4.1 that, in some cases, the activity of metal sulfates is related to their cation electronegativity X_i. In fact, quite good correlations have been obtained between the X_i of metal sulfates and their activity for the hydration of propylene, the polymerization of

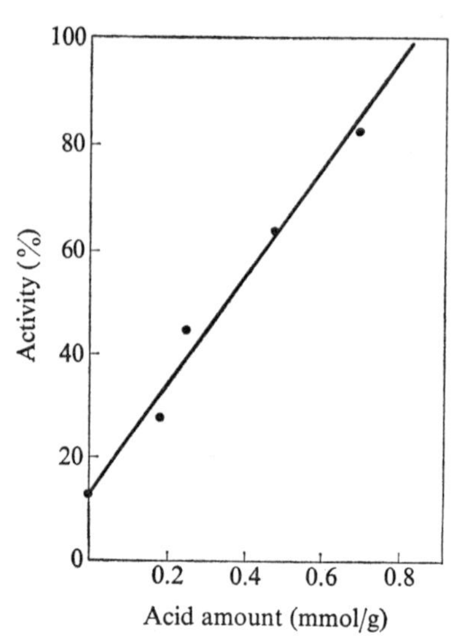

Fig. 5–13 Ethylene oxide polymerization activity *vs.* acid amount at $H_0 \leqq +3.3$ for zirconium phosphate catalysts

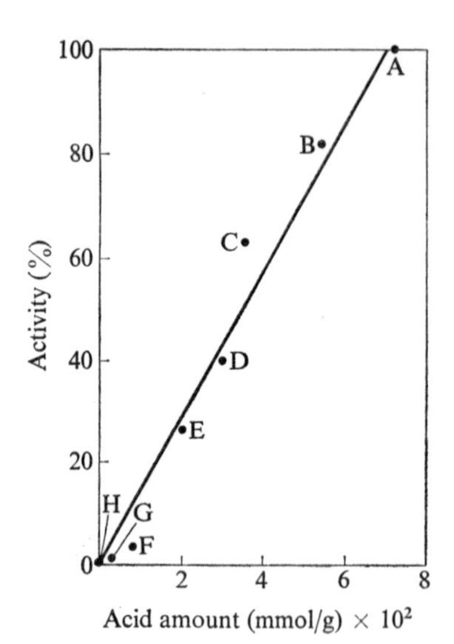

Fig. 5–14 Isopropyl alcohol dehydration activity at 225 °C (measured in terms of propylene yield after 20 min) *vs.* acid amount at $-3 < H_0 \leqq +1.5$ for various metal phosphates
A, B-phosphate (calcined at 280 °C)
B, Fe(III) (260 °C) C, Cr(III) (250 °C)
D, Ti(IV) (280 °C) E, Cu(II) (250 °C)
F, Ni (250 °C) G, Al (250 °C) Zn and Mg (240 °C)
H, Ca and Na (240 °C)

acetaldehyde, the decomposition of formic acid and the exchange reaction of deuterated acetone with water.[46,47] The catalytic activities of metal sulfates for isobutylene polymerization were also found to correlate with X_i (Fig. 5–15).[46] The parameter X_i seems to be useful for predicting the acid amount and the catalytic activity of the hydrated surfaces.

Among mounted acids, phosphoric acid mounted variously on diatomaceous earths, silica gel, quartz sand, etc. has been extensively used as an acid catalyst for the polymerization of olefins, the synthesis of isoprene from 4,4-dimethyl-1,3-dioxane, the alkylation of certain aromatic compounds, etc. The catalytic activity of solid phosphoric acid depends largely upon the kind of carrier and the heat-treatment. As

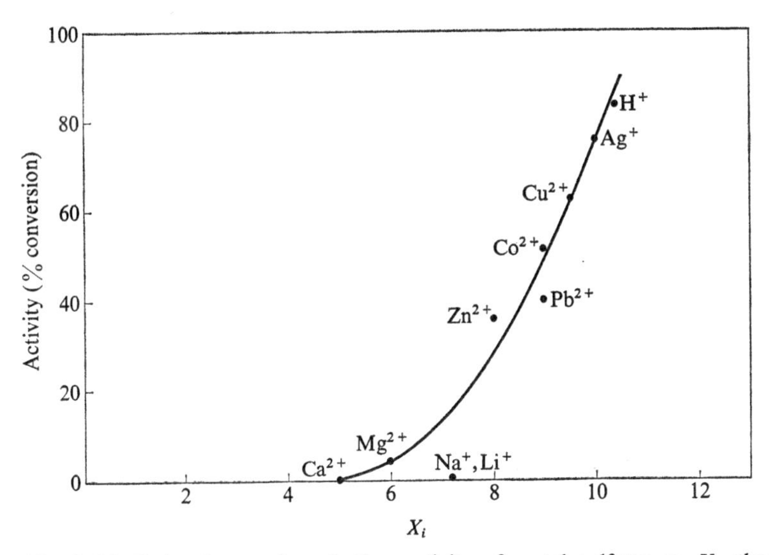

Fig. 5–15 Isobutylene polymerization activity of metal sulfates *vs.* X_i, the metal cation electronegativity

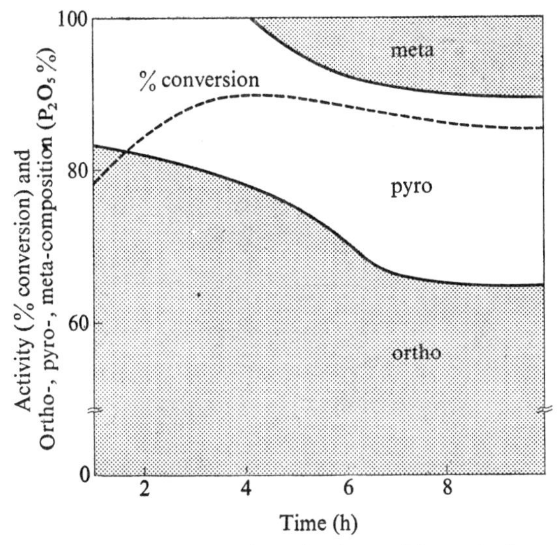

Fig. 5–16 Propylene polymerization activity and phosphoric acid composition *vs.* duration of heat-treatment at 200 °C

illustrated in Fig. 5–16, the activity of phosphoric acid mounted on Makkari diatomaceous earth in the polymerization of propylene increases with the duration of catalyst heat-treatment at 200 °C. The maximum activity, 90% conversion, is attained after 4 h heat-treatment, but gradually decreases as metaphosphoric acid is formed.[48] The activity seems to be intimately linked with the ratio of the compositions of ortho-, pyro- and metaphosphoric acids. Solid phosphoric acid had been thought to lose its acidic properties and activity when heat-treated at high temperatures (in general above 300 °C)[49] but recently Mitsutani and Hamamoto[50] found that the catalysts still show quite high acidity even after heat-treatment at 700~1,200 °C (*cf.* Fig. 4–34 in 4.5.2). They also found that the activity in the isomerization of 1-butene to isobutylene (see the open circles in Fig. 5–17), the dealkylation

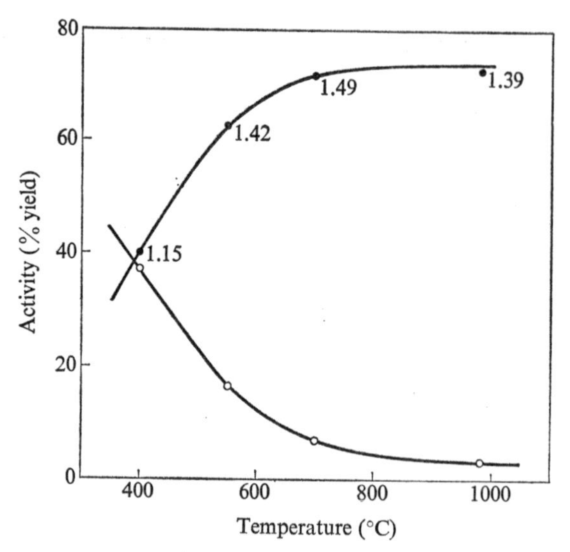

Fig. 5–17 1-Butene isomerization activity in terms of the yield of isobutylene and 2-butene over solid phosphoric acid *vs.* temperature of heat-treatment
(●) 2-butene yield (○) isobutylene yield
The figures refer to *trans*/*cis* ratios

of cumene and the polymerization of propylene each correlate well with the amount of acid (*cf.* Fig. 4–34 and Fig. 5–17). On the other hand, they observed that the activity of the catalyst heat-treated at high temperature in the double-bond isomerization of 1-butene (see the solid circles in Fig. 5–17), the dehydration of isopropanol and the decomposition of 4,4-dimethyl-1,3-dioxane is remarkably higher than that of the catalyst heat-treated at lower temperatures.[51] The decomposition per cent of 4,4-dimethyl-1,3-dioxane over catalysts heat-treated at various temperatures is shown in Table 5–3, where the catalyst selectivity (conversion percent to isoprene) is also given.

TABLE 5–3 Activity and selectivity of solid phosphoric acid heat-treated at various temperatures for the decomposition of 4,4-dimethyl-1,3-dioxane at 200°C

Temperature of heat-treatment (°C)	Activity (%)	Selectivity (%)
100	72.9	60.5
210	70.2	58.9
440	81.9	56.4
570	92.6	61.4
750	95.2	81.2
980	89.9	83.9
1,100	76.6	88.4

5.1.2 Correlation between acid strength and catalytic activity of acid sites

Various examples of correlations between acid amount for sites having a *certain acid strength* and catalytic activity have been given in the preceding section. Next, we shall see how the activity of solid acid catalysts depends upon the acid strength. As mentioned in 5.1.1, the propylene polymerization rate increases with increase in the amount of acid at $H_0 \leqq +1.5$, regardless of the type of acid site.[37] On the basis of an analysis of the reaction products, the predominant reaction at the stationary stage was ascertained to be the trimerization of propylene, which seems to occur on moderately strong acid sites.[52] Yoneda calculated a set of rates V_i of oligomerization per unit (μmol) amount of acid for each of several ranges (i=1, 2, 3, etc.) of acid strength H_0. The results are shown in Fig. 5–18, where the dotted lines indicate standard deviations. The figure gives apparent confirmation of the major role played by acid sites at $H_0 \leqq +1.5$ (i.e. i=4 and 5) in catalytic activity. Similar rate constants for the appropriate range of H_0 were found to increase with the mean acid strength of the range both for the depolymerization of paraldehyde over nickel sulfate[53] and the isomerization of *o*-xylene over silica-alumina,[54] as shown in Fig. 5–19. Conversely, the regional energy of activation decreases with the mean regional acid strength $(H_0)_i$. MacIver *et al.* have also calculated the regional rates for

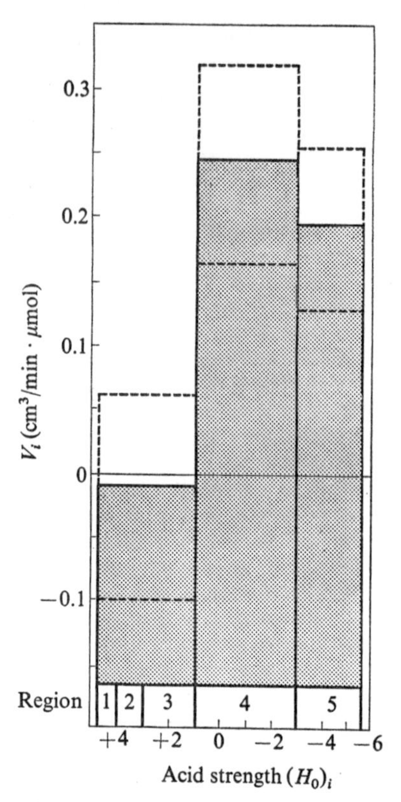

Fig. 5–18 Regional rates V_i of propylene oligomerization over nickel sulfate *vs.* acid strength (regions 1+2+3, 4, and 5)

olefin isomerizations over $SiO_2 \cdot Al_2O_3$ and $SiO_2 \cdot MgO$ catalysts, but the results showed that the distribution in catalytic activity does not appear to parallel the acid strength distribution.[55]

Clear dependence of catalytic activity upon distribution of acid strength is shown in Table 5–4, where the acid strength was determined by gas chromatography.[56] This table indicates that the products from seven compounds over the fresh catalyst were all very similar. Since they were predominantly isobutane, isobutylene, isopentenes, propylene, *n*-butenes and C_6-hydrocarbons, they may all be considered as polymerization and cracking products from butenes, which are presumably the primary products from these reactants. Dehydration of *tert*-butanol takes place very readily even on the weaker acid sites, and the number of sites active in this reaction amounted to more than 0.53 mmol/

TABLE 5–4 Catalytic activity and acid strength of poisoned silica-alumina

Amount of adsorbed pyridine (mmol/g)		Reaction				
		Dehydration	Cracking (A)	Double-bond migration and *trans-cis* isomerization	Cracking (B)	Skeletal isomerization
		tert-butanol to butenes	Diisobutylene to butenes and others	*n*–butenes to its equilibrium position	Dealkylation of *tert*-butylbenzene	isobutylene to *n*-butenes
0	Exceedingly strong acid sites exist	Reaction products from these compounds were very similar, mainly i-C_4, i-C_4', i-C_5, C_3' and other C_4				
0.053	Moderately strong acid sites and weak acid sites	100%	100%	100%	1%	trace
0.106		100	100	100	trace	0
0.149		100	22	1~10	trace	0
0.289	Weak acid sites only	100	trace	trace	0	0
0.415		100	trace	trace	0	0
0.531		12	0	0	0	0

i-C_4: iso-saturated hydrocarbons (carbon number: 4)
i-C_4': isoolefins (carbon number : 4)
i-C_5: iso-saturated hydrocarbons (carbon number: 5)
C_3': *n*-olefins (carbon number: 3)
C_4: *n*-saturated hydrocarbons (carbon number: 4)

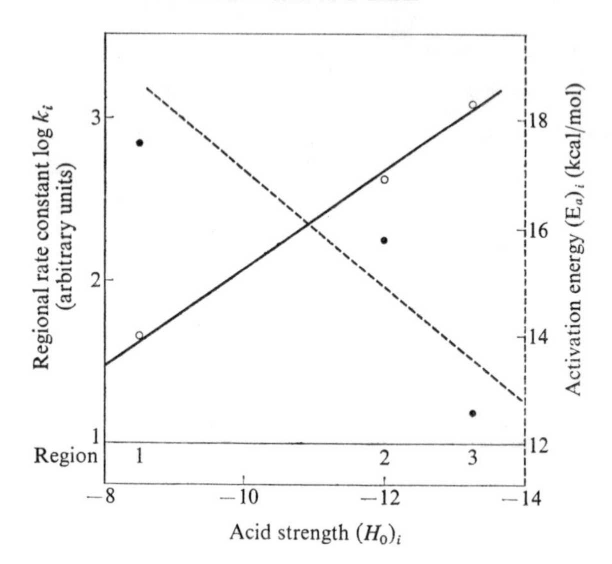

Fig. 5–19 *o*-Xylene isomerization regional rate constants and energies of activation at 500 °C *vs.* mean regional acid strength $(H_0)_i$ of silica-alumina (—○—) rate constants, (--●--) activation energies

g at the temperature of these observations. Further reactions, such as polymerization or cracking of the olefins formed, proceed only on the very strongest acid sites. Dealkylation of *tert*-butylbenzene and skeletal isomerization of isobutylene require exceedingly strong acid sites, and there exist no more than 0.05 mmol/g of such active sites on the surface. Active sites for the cracking of diisobutylene and double-bond migration and *cis-trans* isomerization of *n*-butenes are present to the extent of about 0.15 mmol/g; on the very strongest sites they undergo still further reactions.

The influence of the acid strength of solid catalysts on their activity per unit amount of acid in the dimerization of propylene, isobutylene and the dehydration of isopropyl alcohol has been studied by Dzisko.[25] As may be seen from Table 5–5, the rates of the three reactions do change with the acid strength of the catalysts. The discrepancy between the catalytic activity and the acid strength as measured for $SiO_2 \cdot Al_2O_3$ is apparently due to the fact that no indicators with pK_a less than −8.2 were used. In fact, the strength of the acid sites on some silica-aluminas has recently been shown to be greater than that of an indicator (2,4-dinitrotoluene) with a pK_a of −12.8.[52] The precise character of the dependence of reaction rate on acid strength varies from reaction to reaction: the dependence of the dimerization rate of propylene on acid strength is, for instance, much more marked than that for the dimerization of isobutylene. For each of the three reactions in Table 5–5, however, the dependence of the catalytic activity on the acid strength is less pronounced than for homogeneous acid-catalyzed reactions. The coefficient α in Hammett's equation is only $0.15 \sim 0.6$, compared with $1 \sim 2$ for homogeneous reactions.

TABLE 5–5 Correlation of activity with acid strength (logarithm of reaction rate constants)

Composition of catalysts	pK_a	Dehydration of isopropyl alcohol (l/mmol · h)	Dimerization of propylene (l/mol · h)	Dimerization of isobutylene (l/mol · h)
$Al_2O_3 \cdot SiO_2$	−8.2	3.75	4.1	—
$ZrO_2 \cdot SiO_2$	−8.2	2.25	2.75	4.0
Phosphorous acid on silica gel	−5.6	1.47	1.6	3.6
$MgO \cdot SiO_2$	−3.0	1.25	—	—
Metaphosphorous acid on silica gel	−3.0	0.55	0.0	3.3

Topchieva *et al.* have shown that the activity of alumina in the isomerization of allylbenzene to propenylbenzene at 300 °C does not depend upon structural changes accompanying the appearance of a new phase as a result of heat-treatment.[19] That is to say, the acid sites are identical with the active sites in this reaction, acid sites of all different strengths participating in the reaction, with the stronger sites showing the greater catalytic activity. Pines and Haag found that a mixture of butenes was formed in the decomposition of *n*-butyl alcohol on adulterated alumina, the yield of 1-butene being considerably increased by the addition of small quantities of sodium oxide.[12] From this it was concluded that the γ-alumina surface has acid sites of a range of different strengths, both strong and weak sites being active in dehydration, and only the strong sites active in isomerization. Moreover, the active surface centres of γ- and θ-alumina, prepared by calcining boehmite and bayerite, display identical acidic properties, and the same sites are active for both alcohol dehydration and 1-butene isomerization.[57] Hirota *et al.* have noted that the rate of olefin polymerization over alumina catalysts having strong acid sites drops markedly as the reaction proceeds, but that the reduction is not nearly so marked where the strong acid sites of the alumina have previously been poisoned with silver.[58] Clark *et al.* have studied the relationship between the differential surface entropy (a measure of the mobility of adsorbed molecules) for ammonia adsorbed on a series of $SiO_2 \cdot Al_2O_3$ catalysts, and their catalytic activity for various acid-catalyzed reactions.[59] They concluded that the cracking of *n*-octane, the polymerization of propylene, the isomerization of *o*-xylene and the hydrogen transfer between 1-butene and decalin take place on relatively weak acid sites, while the hydrogen exchange reaction takes place on comparatively strong sites. Hsieh also discusses the correlation of the cracking activity of $SiO_2 \cdot Al_2O_3$ with the differential heat of adsorption of ammonia (an index of acid strength) shown in Fig. 2–4 (*cf.* 2.1.3).[60] Evidence for the existence of more than one type of catalytically different active site on a given solid acid surface has been obtained by Brouwer, who observed a selective aging effect accompanying the reaction of *cis*-2-butene over γ-alumina,[61] by Pines *et al.*, who studied olefin reactions over various alumina catalysts (*cf.* 2.1.4),[12] and by MacIver *et al.*, who noted that in the reaction of 1-pentene over alumina catalysts dried at different temperatures, the double-bond and skeletal isomerization activities did not seem to be in conformity.[62] For zeolite catalysts, Norton reports that the activity of calcium ion exchanged zeolite in propylene polymerization and its acid strength are both higher than those of the unexchanged zeolite.[63] For $Al_2O_3 \cdot B_2O_3$ catalysts, an experiment involving poisoning by ammonia has shown that there is an optimum strength for the acid sites if they are to be effective in the cracking of cumene.[64]

Metal sulfate catalysts have been the object of extensive studies to identify those strengths of acid site which have the highest catalytic activity in various reactions. This is because they have, in general, a wide range of acid strengths (see Fig. 4–24 and 4–25), with a preponderance of acid centres of moderate strength, especially as compared with $SiO_2 \cdot Al_2O_3$, Al_2O_3, $AlCl_3$, etc., which have high acid strengths, and with TiO_2 and ZnS, whose acid strengths are very low. Table 5–6 summarizes the acid strengths of metal sulfates and phosphates which are effective in various reactions. The range of effective acid strengths was determined by observing the correlation between the activity and the amount of acid *at various acid strengths*, or by determining the activity with catalysts whose acid sites had been selectively poisoned with an indicator having a suitable pK_a value. In the majority of the reactions listed it is clearly the comparatively weak acid sites which are catalytically active. In the case of propylene polymerization, all sites with $H_0 \leqq +1.5$ are active (see again the discussion in 5.1.1), whereas in the hydration of propylene those sites with intermediate strengths are clearly the most active.

Metal sulfates are also effective catalysts for the formation of forma-

TABLE 5–6 Effective acid strength of metal sulfate catalysts for various reactions

Reactions	Effective strength	Reference No.
Beckmann rearrangement of cyclohexanone oxime	$H_0 \leqq +4.0$	24
Disproportionation of fluorochloromethane	$H_0 \leqq +3.3$	40
Isomerization of α-pinene	$H_0 \leqq +3.3$	66
Polymerization of aldehydes	$H_0 \leqq +3.3$	35
Condensation of glucose with acetone	$H_0 \leqq +3.3$	67
Esterification of phthalic acid	$H_0 \leqq +3.3$	36
Polymerization of propylene	$H_0 \leqq +1.5$	37
Depolymerization of paraldehyde	$H_0 \leqq -3$	28, 29
Hydration of propylene	$-3 < H_0 \leqq +1.5$	41
Dehydration of isopropanol	$-3 < H_0 \leqq +1.5$	45
Methanol synthesis from carbon dioxide and hydrogen	$+1.5 < H_0 \leqq +3.3$	27
Polymerization of isobutylvinylether	$-5.6 < H_0 \leqq -3$	42

lin from methylene chloride,[65] but the reaction is accompanied by some side reactions when silica-alumina is used as the catalyst. This can be explained on the assumption that the high acid strength of silica-alumina is responsible for the side reactions. On the other hand, there are solid acids with only weak sites at $H_0 > +3.3$ such as titanium oxide and zinc sulfide, which do not show any catalytic activity for the polymerization of propylene,[37] the isomerization of pinene,[66] or the condensation of glucose with acetone.[67] It is characteristic of metal sulfate catalysts that the acid sites of moderate strength suitable for a particular reaction may readily be formed by heat-treatment.[30]

5.1.3 Correlation between type of acid site and catalytic activity

The correlations which have so far been cited between catalytic activity and the amount and strength of acid sites do not, of course, differentiate between Brønsted and Lewis acid sites, for the analytical methods which have been used (amine titration using indicators and many of the other methods described in 2.1 and 2.2) are effective for both types, *cf.* 2.2.1. A catalyst may well possess only Brønsted (or Lewis) sites, in which case the acid properties will naturally be exclusively of Brønsted (or Lewis) type. Even so, the above methods do not enable us to distinguish between them. For knowledge of the correlations between catalytic activity and either Brønsted or Lewis acidity we must turn to the specific methods described in 2.3.

Shephard *et al.* studied the polymerization of propylene on a silica-alumina catalyst and found that the activity was drastically curtailed when surface hydrogen atoms of the catalyst were exchanged with sodium ions, suggesting that Brønsted acid sites are essential to the reaction.[68] This conclusion is in harmony with the results of Holm, Bailey and Clark, who showed that the polymerizing activity of a series of silica-alumina catalysts correlated well with their Brønsted acidity as measured by the ammonium ion exchange method (*cf.* 2.3.1), but not with their total acidity measured by the amine titration method.[69] Fig. 5–20

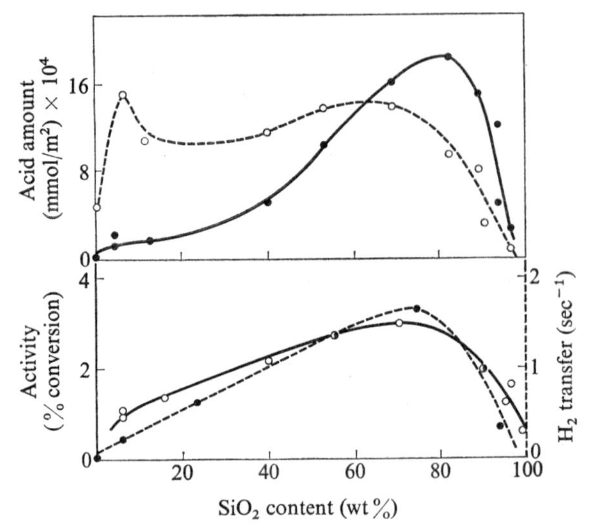

Fig. 5–20 Acid amount and catalytic activity *vs.* SiO_2 content for a series of $SiO_2 \cdot Al_2O_3$ catalysts (—●—) Brønsted acid amount, (--○--) total acid amount (amine titration method), (—○—) propylene polymerization, (--●--) hydrogen transfer

gives their results. Shiba *et al.* have also demonstrated a clear connection between the polymerization activity of silica-aluminas and Brønsted acidity, the latter being derived from the difference between the total acidity and the Lewis acidity (*cf.* 4.2.1).[64] On the other hand, the fact that the greater proportion of olefins is formed from propylene on silica-alumina treated with perylene, which is thought to be adsorbed only on Lewis sites and not to any appreciable extent on Brønsted sites, suggests that Lewis sites may be responsible for those reactions which supply the necessary hydrogen for the conversion of olefins and carbonium ions to saturated compounds.[68] This conclusion is in conformity with the hypothesis that Lewis sites are active in the accumulation of aromatic compounds and coke, which was substantiated by the ESR investigations of Rooney and Pink.[70] MacIver *et al.* observed that perylene, which specifically poisons Lewis acid sites, had no effect on the activity or the selectivity of $SiO_2 \cdot Al_2O_3$ and $SiO_2 \cdot MgO$ for the skeletal isomerization of 3,3-dimethyl-1-butene and the double-bond isomerization of 1-pentene.[55] This suggests that the Lewis sites are inactive for these isomerization reactions.

Brouwer has reported that fluorided γ-alumina, $SiO_2 \cdot BF_3$, and silica-alumina treated with sodium acetate all oxidize perylene to the cation radical, while γ-alumina, silica and fluorided silica all fail to do so.[71] Sodium acetate drastically reduces the cumene-cracking activity of silica-alumina, but has only a minimal effect on the oxidation of perylene. This observation suggests that Brønsted sites are responsible for cumene-cracking activity, while the oxidation of perylene to the cation radical takes place on the Lewis sites.[71] Brouwer has also pointed out that the oxidation of perylene apparently occurs on the same centres as the oxidation of triphenylmethane to the triphenylcarbonium ion. It is also known that perylene and anthracene are oxidized by BF_3, a typical Lewis acid, and 98% sulfuric acid, but not by Brønsted acids such as HF in the absence of molecular oxygen.[72] Hirschler and Hudson suggest that in the reaction of perylene with silica-alumina to form the perylene cation radical, Brønsted sites catalyze the reaction with some form of chemisorbed oxygen which acts as the electron acceptor.[73]

Shiba *et al.* have shown that the activity of silica-alumina for the polymerization of isobutylene and the cracking of cumene has a close correlation with the amount of Brønsted acid,[64,74] while that for the decomposition of isobutane with Lewis sites is as shown in Fig. 5–21, where the amount of Lewis acid was determined by Leftin and Hall's method (*cf.* 2.3.2). Despite the questionable reliability of this method the activity was found to be almost exactly proportional to the amount of Lewis acid for two series of silica-alumina samples prepared by quite different methods. A more reliable method for the measurement of Brønsted or Lewis acidity is that based on the infrared spectrum of chemisorbed pyridine (*cf.* 2.3.3). Fig. 5–22 shows the linear relationship which holds between the amount of Brønsted acid and the catalytic activity for *o*-xylene isomerization as determined by this method.[75]

The activity of a partially halogenated porous glass catalyst for the cracking of cumene was shown to be due to true Brønsted (and not Lewis)

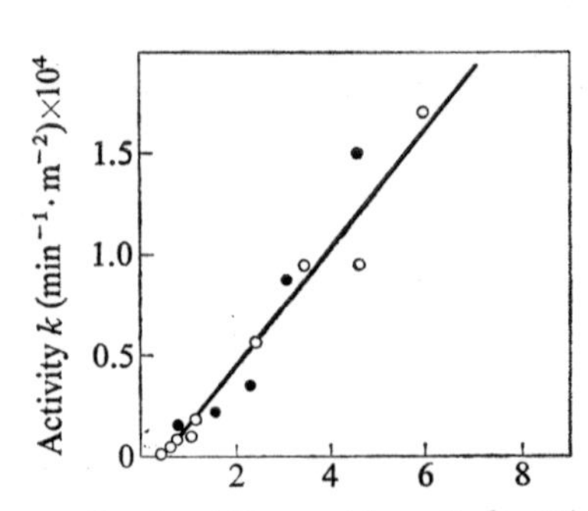

Fig. 5–21 Isobutane decomposition activity *vs.* Lewis acid amount for various $SiO_2 \cdot Al_2O_3$ catalysts
(●) alumina fraction prepared from aluminium isopropoxide
(○) from aluminium nitrate

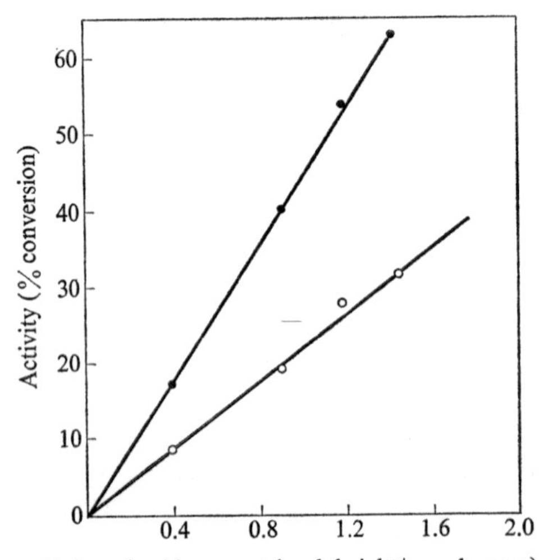

Fig. 5–22 *o*-Xylene isomerization activity *vs.* amount of Brønsted acid for $SiO_2 \cdot Al_2O_3$
(●) heat-treated at 500 °C, (○) 425 °C

acid sites[76] (*cf.* diagram A, page 54). Both the amount and strength of Brønsted acid on silica gel is known to increase remarkably upon adsorption of a halogen.[77] Recently, Antipina *et al.* have clearly shown the correlation between the amount of Brønsted acid on fluorided alumina and aluminium hydroxyfluorides and their cumene-cracking activity (see Fig. 5–23).[78] Both the activity and the Brønsted acidity, as measured by *n*-butylamine titration with an H_R indicator (aryl-methanol) of $pK_a=-13.3$, increase with increasing catalyst fluorine content.

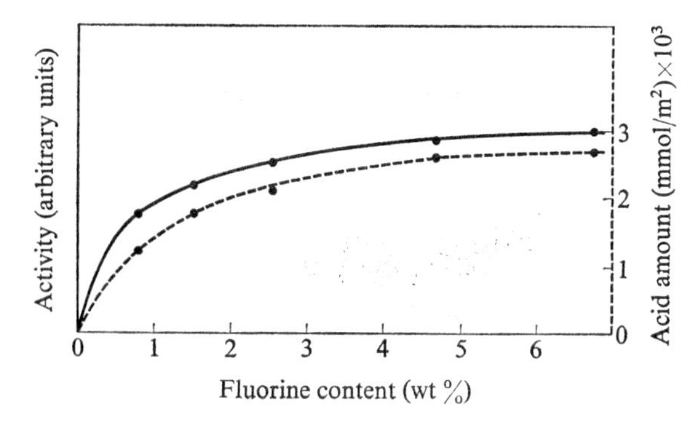

Fig. 5–23 Cumene cracking activity and amount of Brønsted acid at $H_0 \leqq -13.3$ *vs.* fluorine content in fluorided alumina catalysts (—●—) activity, (--●--) acid amount

The isomerization of 1-butene on alumina is reported to be catalyzed by Lewis acid sites.[79] In the case of $SiO_2 \cdot Al_2O_3$ and $NiO \cdot SiO_2$, Ozaki and Kimura hold that the Lewis acid sites can be effective for isomerization, a proton being donated by the olefin molecule chemisorbed on the Lewis acid site.[80] Thus the Lewis sites can, under some circumstances, produce effects similar to Brønsted sites.

It was mentioned in Chapter 4 that both types of acid sites, or either one of them, can exist on cation exchanged zeolites, and that some of the Lewis acid sites are thought to be converted into Brønsted sites with the addition of water. Brønsted type acid at strengths greater than $H_0= +1.5$ is active in propylene polymerization, and these converted Lewis sites are similarly active, with the exception of La-zeolite.[20] There is a difference, however, in their activities for propylene and ethylene polymerization which cannot be explained simply in terms of Brønsted acidity. Also in the double-bond isomerization of 1-butene, although both types of the acid are active, there appears to be a difference in their activities.[20] Ward has found that the cumene-cracking activity of cation exchanged zeolite Y increases with increasing Brønsted acidity determined by the infrared method, see Fig. 5–24.[81] According to Topchieva *et al.*, the active centres of both decationized and cationized zeolites in cumene-cracking are protonized OH groups.[82] In hexane conversion, an increase in activity accompanies any increase in Brønsted acidity; those zeolites having cations of smaller ionic radius are also correspondingly more active (see Fig. 5–25).[81] The high activity of zeolite and mordenite (NH_4^+ form) heat-treated at 600～700 °C in the disproportionation of toluene and the reaction of *n*-paraffins is also assumed to be due to Brønsted sites.[83]

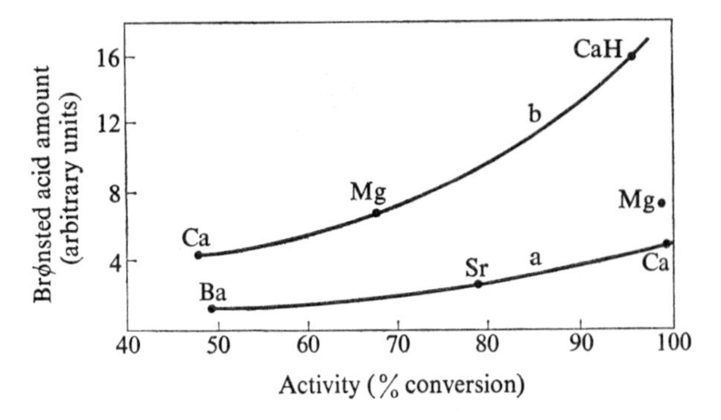

Fig. 5–24 Cumene cracking activity of various cation exchanged zeolites Y *vs.* amount of Brønsted acid
a: cracking at 500 °C, b: 260 °C

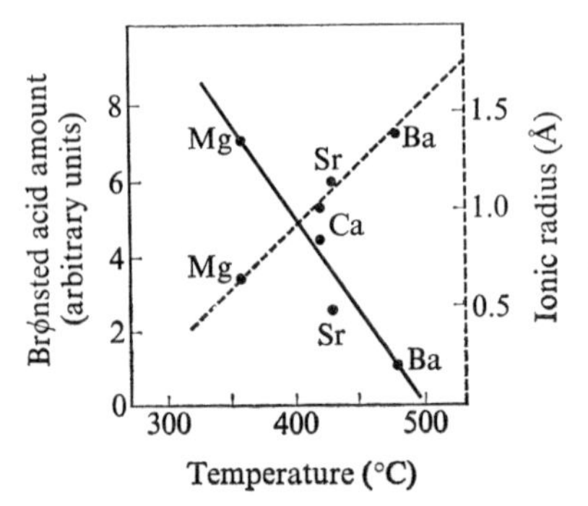

Fig. 5–25 Brønsted acid amount and ionic radius *vs.* temperature which gives 5% hexane conversion
(—●—) acid amount
(--●--) ionic radius

In the case of alumina-boria catalysts, the activity for the disproportionation of toluene was found to correlate well with the Brønsted acidity, but not significantly with either the Lewis acidity or the total acidity, as illustrated in Fig. 5–26.[84] Fig. 5–27 shows the relationship between the catalytic activity per unit amount of Brønsted acid and the boria content. The increase in activity with increasing boria content is thought to be due to the change in the strength of the Brønsted acid sites.[84] In the case of silica-magnesia, Bremer and Steinberg found by infrared

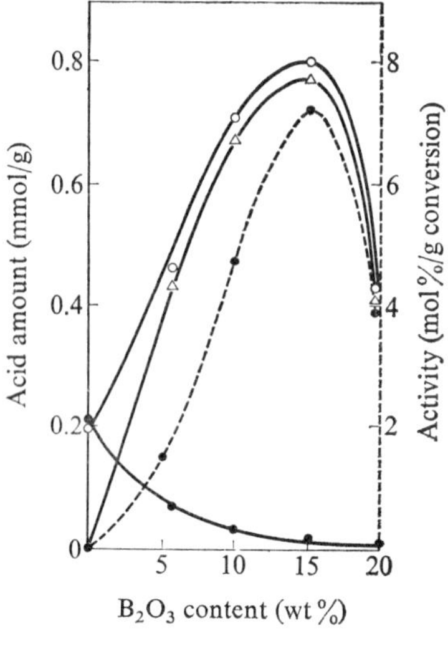

Fig. 5–26 Toluene conversion activity and acid amount *vs.* B_2O_3 content of $Al_2O_3 \cdot B_2O_3$
(—○—) total acid amount
(—△—) Brønsted acid amount
(—●—) Lewis acid amount
(--●--) toluene conversion

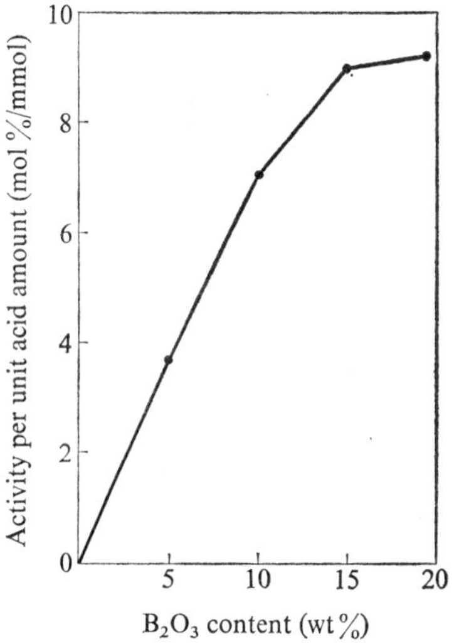

Fig. 5–27 Toluene disproportionation activity per unit amount of Brønsted acid *vs.* B_2O_3 content for various $Al_2O_3 \cdot B_2O_3$ catalysts

measurements that moderately strong Lewis type acid sites and weakly acidic isolated OH groups exist on the surface, and suggested that the catalyst acts as a bifunctional acid catalyst for the dehydration of isopropanol.[85]

It was pointed out in 5.1.1 that the maximum rate of depolymerization of paraldehyde when catalyzed by solid nickel sulfate is considered to coincide with the maximum amount of (L+B) acid as determined by amine titration (*cf.* Fig. 4–25). This reaction is known to be catalyzed not only by Brønsted acid (H_2SO_4, CCl_3COOH, etc.),[32] but also by Lewis acid ($TiCl_4$, $AlCl_3$, $SnCl_4$, etc.).[86] However, the maximum rate of isomerization of α-pinene to camphene with the same catalyst coincides not with the maximum of the total amount of acid (which appears on nickel sulfate heat-treated at 350 °C), but with the maximum for the amount of Brønsted acid (heat-treated at 250 °C) which was measured by the infrared method, see Fig. 5–28 (*cf.* Fig. 4–30). A similar phenomenon is also observed in the isomerization of 1-butene to 2-butene by nickel sulfate catalysts mounted on silica.[80] Again, in the conversion of methylene chloride to formaldehyde, the maximum catalytic activity was obtained when the nickel sulfate was heat-treated at 400 °C.[65] In

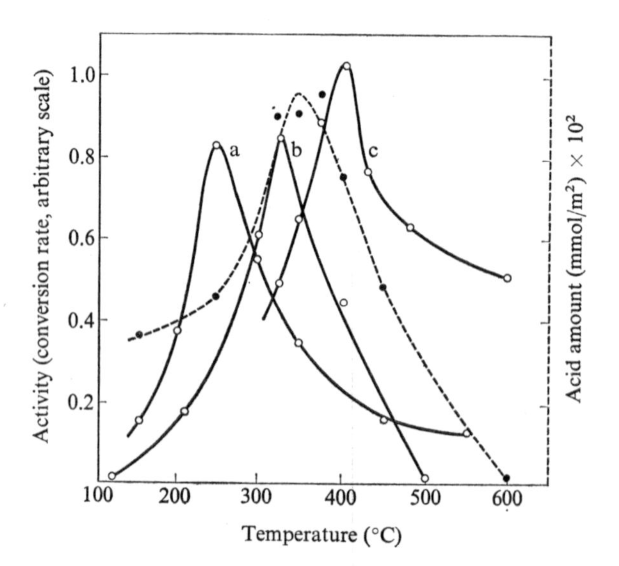

Fig. 5–28 Catalytic activities and amount of acid (B+L) on nickel sulfate *vs.* temperature of heat-treatment
a: isomerization of α-pinene to camphene
b: depolymerization of paraldehyde
c: conversion of methylene chloride to formaldehyde
The dotted line gives acid amount at $H_0 \leqq +3.3$.

this case the maximum coincides with that of the Lewis acidity (again *cf.* Fig. 4–30 and 5–28). The conclusion is, therefore, that the hydrolysis of methylene chloride at temperatures above 300 °C is catalyzed by Lewis acid, whereas the isomerizations of α-pinene and 1-butene are catalyzed by Brønsted acid sites. Confirmation of the role of Lewis acid sites came from employing the series of catalysts listed in Table 5–7.

TABLE 5–7 Acid properties of various solid acids and their activity for hydrolysis of CH_2Cl_2 at 350 °C

Catalysts	Acid sites				Activity
	$H_0 \leqq +4.8$	$H_0 \leqq +4.0$	$H_0 \leqq +3.3$	Type	
$NiSO_4 \cdot xH_2O$	○	○	○	B+L	○
ZnS	○	○	×	B† +L	○
$SiO_2 \cdot Al_2O_3$ ($H_0 \leqq -8.2$)	○	○	○	B+L	low
H_3PO_4/Diatomaceous earth	○	○	○	B	×
HCl/Diatomaceous earth	○	○	○	B	×
NiO	×	×	×		×
SiO_2	○	×	×		×
Diatomaceous earth		×	×		×

○ : Present × : Absent † : Unconfirmed

Of these, the Brønsted acids such as phosphoric acid and hydrochloric acid mounted on diatomaceous earth were inactive, while all catalysts containing Lewis acid sites with strengths $H_0 \leqq +4.0$ were active.

The methylene chloride reaction is known to be catalyzed by hydroxyl ions, but not by sulfuric acid in solution[87] or hydrogen chloride in the vapour phase:[88] the mechanism is considered to be a bimolecular nucleophilic substitution reaction involving the hydroxyl ion.[87] However, when the reaction is catalyzed by Lewis acids such as metal sulfates, zinc sulfide, metal ions[89] etc., the reaction rate is thought to be governed by the extraction of chloride ions from the methylene chloride by the Lewis acid to form carbonium ions $C^{\oplus}H_2Cl$.

5.1.4 Selectivity and characteristics of solid acid catalysts

The selectivity of a solid acid catalyst is influenced by its acidic properties and many other factors such as its geometric structure (particularly pore structure), the distribution of basic sites if present, the polarity of the surface, etc. We have already encountered several examples where the selectivity is governed by the strength and the type of acid sites. In the reaction of phenol with methanol, the products are controlled to considerable extent by the acid strength of the catalyst. With a strong solid acid ($SiO_2 \cdot Al_2O_3$, etc.), in addition to alkylation there can occur simultaneously methanol dehydration, cresol isomerization, etc., or the dealkylation of xylenol. With a catalyst of moderate strength (solid phosphoric acid, some types of alumina, etc.), only the alkylation itself is observed.[90] Recently, several catalysts including $4MgCO_3 \cdot Mg(OH)_2 \cdot 4H_2O$,[91] $Ce_2O \cdot MnO \cdot MgO$[92] and $Fe_2O_3 \cdot ZnO$[93] were found to show extremely high selectivity for the formation of 2,6-xylenol, although the correlation between selectivity and the acidic and/or basic properties of these catalysts has not yet been investigated.

Rearrangements of olefins which proceed via primary carbonium ions are believed to progress with any great rapidity only on relatively strong acid sites, whereas those involving more stable secondary and tertiary carbonium ions can occur equally well on both strong and weak sites.[12]

MacIver *et al.* have studied the effect of varying degrees of ammonia chemisorption on the activity and on two specific selectivities of silica-alumina catalysts for olefin isomerizations.[55] Table 5–8 lists the average amounts of chemisorbed ammonia still retained by the catalyst after each kinetic run, the amount of carbon deposited, and Type I selectivity (the ratio of two products obtained via two parallel paths from the same

TABLE 5–8 Type I selectivity, S (I), of silica-alumina

Reactants	Surface (evacuation temp.)	Fraction NH_3 after run	Carbon (wt%)	140°C		155°C	
				S (I)	Conv. (%)	S (I)	Conv. (%)
DMB	unpoisoned	—	4.65	4.0±0.1	17~40	3.5±0.1	17~45
DMB	NH_3 (400°)	—	2.03±0.10	4.0±0.0	16~25	3.7±0.2	19~27
DMB	NH_3 (300°)	0.64	2.43±0.71	3.9±0.2	8~13	3.6±0.2	10~14
DMB	NH_3 (200°)	0.94±0.06	0.66±0.07	3.9±0.0	2~4	3.5±0.2	4~6
DMB	Perylene	—	—	3.8±0.1	20~30	3.7±0.2	20~35
1-Pentene	unpoisoned	—	6.45±0.40	1.02±0.04	20~32	1.08±0.08	23~28
1-Pentene	NH_3 (400°)	0.78±0.06	3.13	0.98	8	1.00	13
1-Pentene	NH_3 (300°)	0.88±0.12	2.16±0.01	0.98±0.01	3~5	0.98±0.01	5~6
1-Pentene	NH_3 (200°)	0.54±0.00	0.50±0.08	1.00±0.04	2~3	1.02±0.02	3~4
1-Pentene	Perylene	—	5.90±1.75	1.18±0.05	15~18	1.20±0.04	16~21

For DMB (3,3-dimethyl-1-butene), S (I) = (2,3-dimethyl-2-butene)/(2,3-dimethyl-1-butene)
For 1-pentene, S (I) = (*trans*-2-pentene)/(*cis*-2-pentene)

reactant). The data show that poisoning silica-alumina with ammonia has no appreciable effect on the Type I selectivity for either reactant. However, a form of Type II selectivity (the ratio of the activities of a catalyst for carbon production from two different reactants — 1-pentene and 3,3-dimethyl-1-butene in this case) is evident. Sodium hydroxide poisoning has been observed by Dzisko *et al.* to have a similar effect on the catalytic activity and selectivity of alumina in the dehydration of *n*-butyl alcohol.[57] The proportion of 1-butene in the reaction products was observed to be almost constant despite the marked variations in the extent of poisoning and in the catalytic activity. Gerberich and Hall have found indications of different selectivities for alumina and silica-alumina.[94] Their observations reveal that whereas the isomerization rate of 1-butene increases over silica-alumina with increasing hydroxyl content, it decreases over alumina. Alumina also differs from silica-alumina in having a higher ratio of *cis*- to *trans*-2-butene in the reaction products. Again, the isomerization rates for *cis*-2-butene and 1-butene are approximately the same over alumina, although the latter is converted much more rapidly than the former over silica-alumina.

Mourgues *et al.* studied the activities of silica-alumina catalysts for a number of reactions, and their selectivities for ethanol dehydration, as a function of the amount of alumina which was removed.[95] This led them to assume the existence of two types of active site, A and B: type A, which strongly adsorbs the reactants but not water, and acts selectively for olefin formation and cracking, and type B, which does not strongly adsorb the reactants, is not poisoned by pyridine, and acts selectively for ether formation.

The weaker acid sites of metal sulfates also play a very important role in selectivity. Metal sulfates are found to be highly selective for various reactions, among them the formation of formalin from methylene chloride,[65] the hydration of ethylene,[96] the dehydration of 4-methyl-2-pentanol,[97] etc., especially in comparison with catalysts with very strong acid sites ($SiO_2 \cdot Al_2O_3$, etc.). In the benzylation of substituted benzenes with benzyl chloride over metal sulfates, the positional reactivities of toluene, o_f, m_f, and p_f were found to be 11.2, 1.30 and 23.2 respectively.[98] This positional selectivity means that toluene in the *para* position reacts 23.2 times faster than benzene. On the other hand, the substrate selectivity among the benzene derivatives was as follows: toluene (1.00), benzene (0.12), *o*-xylene (2.5), *m*-xylene (4.1), *p*-xylene (1.8), and bromobenzene (0.019). The rate is strongly dependent upon electron availability as shown by the ρ-value obtained from Brown's $\log p_f - S_f$ relationship: $\rho = -4$ to -5,[99] where S_f is the selectivity factor $[S_f \equiv \log(2 \times \%\ para/\%\ meta)]$. The very fact that the value of

$\log p_f - S_f$ obtained from the $NiSO_4$ catalyzed benzylation reaction fits Brown's relationship is consistent with Brown's postulate that the slowest step in the reaction is the crucial σ-complex formation.[100] Although the nature of the active surface is not known, it can be said that the surface activity of various metal sulfates is virtually identical as revealed by the positional isomer distribution for benzyltoluene, see Table 5–9, where $AlCl_3$ is included for comparison.

TABLE 5–9 Positional selectivity of metal sulfates

Catalysts	*o*	*m*	*p*
$MnSO_4$	39.8	6.8	53.4
$ZnSO_4$	41.8	6.8	51.4
$CuSO_4$	42.4	6.3	51.3
$CoSO_4$	38.3	6.2	55.5
$FeSO_4$	43.2	7.7	49.1
$AlCl_3$	10.2	69.0	20.8

Although the important role of the acidic properties of solid acids in their catalytic activity and selectivity is acknowledged, other factors are also held to be influential. We first consider here, very briefly, the effect of the structure of alumina-boria on its catalytic activity in the disproportionation of toluene. Catalyst prepared from η-alumina shows higher activity than that from χ-alumina, while α-alumina does not give rise to any catalytic activity. In other words, a higher order of crystallinity results in a reduction of the reactivity.[23] It also seems that very pronounced selectivities and activities can result when the basic sites (found on the surfaces of several so-called solid acids) and the acid sites cooperate. This acid-base bifunctional catalysis is discussed in 5.3.

5.2 Solid base catalysis

In this section we first consider those solid base catalysts which have been used for particular reactions, and then proceed to discuss the correlations between basic properties and catalytic activities. In some cases the catalysis is compared with that of solid acids or homogeneous bases, and the mechanism of heterogeneous base catalysis is discussed.

5.2.1 Reactions catalyzed by solid bases

A. Polymerization, isomerization and alkylation. The oxides, carbonates and hydroxides of alkali metals and alkaline earth metals (MgO, CaO, SrO, Na_2CO_3, K_2CO_3, $CaCO_3$, $SrCO_3$, NaOH, $Ca(OH)_2$, etc.) have been found active in the high polymerization of formaldehyde,[101] ethylene oxide,[102-3] propylene oxide,[104-5] lactam[106] and β-propiolactone.[107] It should be noted in passing that the catalytic activity of $Ca(OH)_2$ is less than that of CaO for β-propiolactone under similar reaction conditions.[107] Not all of these substances, however, necessarily act as base catalysts. The polymerization of lactam is thought to be catalyzed by a complex formed from the monomer and the catalyst,[108] and infrared investigations have actually revealed bond formation between alkali metal and monomer nitrogen atoms.[109] In the case of ethylene oxide polymerization by magnesium oxide, the magnesium cation is considered to play an important role as the active centre.[103]

The dimerization of olefins, particularly the synthesis of hexane from propylene, has been studied using solid bases such as Na/K_2CO_3 (reaction temperature 100 °C),[110] KH dispersed in mineral oil (98 ~ 100 °C),[111] and KNH_2 on alumina (200 ~ 206 °C).[112] The reaction is thought to proceed via carbanion intermediates as illustrated below:

$$CH_3{-}CH{=}CH_2 + B^-M^+ \longrightarrow BH + M^+\bar{C}H_2{-}CH{=}CH_2 \quad \text{(B=H, NH}_2\text{, alkyl, allyl, alkenyl etc., M=alkali metals)} \tag{5.7}$$

$$CH_2{=}CH{-}\underset{M^+}{\bar{C}H_2} + \underset{CH_3}{CH}{=}CH_2 \longrightarrow CH_2{=}CH{-}CH_2{-}\underset{CH_3}{CH}{-}\bar{C}H_2M^+ \quad \text{(I)} \tag{5.8}$$

$$CH_2{=}CH{-}\underset{M^+}{\bar{C}H_2} + CH_2{=}\underset{CH_3}{CH} \longrightarrow CH_2{=}CH{-}CH_2{-}CH_2{-}\underset{CH_3}{\bar{C}HM^+} \quad \text{(II)} \tag{5.9}$$

and chiefly,

$$\text{(I)} + CH_3{-}CH{=}CH_2 \longrightarrow \underset{M^+}{\bar{C}H_2}{-}CH{=}CH_2 + CH_2{=}CH{-}CH_2{-}\underset{CH_3}{CH}{-}CH_3 \quad \text{(III)} \tag{5.10}$$

The allyl anion formed according to reaction 5.7 reacts with propylene according to 5.8 or 5.9, but gives (III) as its chief product, due to the stability of primary carbanions (I). Allyl ions formed as in 5.10 also react with propylene to give the dimer. Further reactions of the kind culminating in 5.11 (below) take place to give 4-methyl-2-pentene (VI).

$$\text{(III)} + B^-M^+ \longrightarrow BH + CH_2{=}CH{-}\underset{M^+}{\overset{-}{C}H}{-}\underset{CH_3}{\underset{|}{CH}}{-}CH_3$$

(IV)

$$\text{(IV)} \longrightarrow \underset{M^+}{\overset{-}{C}H_2}{-}CH{=}CH{-}\underset{CH_3}{\underset{|}{CH}}{-}CH_3$$

(V)

$$\text{(V)} + BH \longrightarrow CH_3{-}CH{=}CH{-}\underset{CH_3}{\underset{|}{CH}}{-}CH_3 + B^-M^+ \qquad (5.11)$$

(VI)

That is, basic catalysts are effective in the isomerization of olefins, and the olefin product distribution is characteristic of thermodynamical equilibrium after a sufficiently long contact time.

A solid base like Na/Al_2O_3 which has a large surface area is very active for the isomerization of olefins. For example, equilibrium for the isomerization of 1-pentene is attained in 60 min even at 30 °C, and that for 1-butene is only 0.6 min at 25 °C.[113-4] Calcium oxides prepared from calcium hydroxide either calcined at 500 °C or at 900 °C (with subsequent annealing at 500 °C), or from calcium carbonate calcined at 900 °C, were each found to be active for the double-bond isomerization of 1-hexene.[115] Since bases such as ammonia and pyridine do not deactivate the catalysts, the isomerization is considered to proceed by a mechanism which involves an ionic allylic intermediary. A similar mechanism has been postulated for base-catalyzed isomerization of alkenes.[113] Clark and Finch, on the basis of results from ammonia blocking, H–D exchange and radiochemical experiments using ^{14}C, suggest that the isomerization of 1-butene over magnesium oxide may proceed by an anionic mechanism which is independent of catalyst acidity.[116] It is also known that the stereoselective isomerization of 1-butene to *cis*- and *trans*-2-butene[117] and the double-bond isomerization of limonene[118] are both catalyzed by basic catalysts, including calcium oxide.

Olefin alkylation of aromatic compounds is also catalyzed by solid bases such as Na/Al_2O_3, NaH, K/graphite, etc. It is significant that the α-carbon of aromatic side chains having benzylic hydrogen is alkylated over base catalysts, while aromatic rings are alkylated over acid catalysts. This side chain alkylation catalyzed by solid bases is also thought to proceed via carbanion intermediaries.[113]

B. Condensation, addition, dehydrohalogenation. The condensation of propionaldehyde is catalyzed by basic lithium phosphate, $Ca(OH)_2$, etc.[119] Although in the case of $Ca(OH)_2$ the main reaction is accompanied by a side reaction (the Cannizzaro reaction), basic lithium phosphate is highly selective. Aldol formation from acetaldehyde and diacetone from acetone are both catalyzed by an anion exchange resin (Amberlite IRA–400, OH type). The activities for aldol-type formaldehyde[120] and acetaldehyde[121] condensation over $NaOH/SiO_2$, MgO, CaO, and PbO correlate well with the basic properties of the surfaces, as described below in 5.2.2. MgO, CaO and K_2CO_3 act as catalysts for the formation of β 2-furylacrolein from furfural and acetaldehyde in the gaseous phase,[122] and Li_2CO_3/SiO_2 for the synthesis of acrolein.[123]

$$\text{(furan)-CHO} + CH_3CHO \longrightarrow \text{(furan)-CH=CHCHO}$$

(yield: MgO, 70.0%; CaO, 36.0%; K_2CO_3, 25.6%)

$$CH_3CHO + CH_2O \xrightarrow[350°C]{Li_2CO_3/SiO_2} CH_2{=}CH{-}CHO$$

(yield: 55~53%)

The Knoevenagel reaction[124] and Michael condensation[125] readily take place in the presence of weakly basic Amberlite IR–4B, Dowex–3 and polyvinylpyridine.

Anion exchange resins are also effective for the addition reaction of alcohol to acrylonitrile,[126] hydrogen cyanide to ketone,[127] nitromethane to aldehyde[128] and of epoxide to phenol.[129]

The dehydrochlorination of 1,1,2-trichloroethane is catalyzed by alkaline earth metal oxides.[130] The oxide catalysts CaO and Al_2O_3 are similarly active in the cases of 1,2-dichloro-2,2-diphenylethane[131] and 2,3-dichlorobutane.[132] The correlation between activity in dehydrochlorination and basic catalyst properties is discussed in the following section.

In the reaction of acrolein with ethanol over various metal oxides, the catalytic activity is found to increase with increasing differences in electronegativities, an indication that the reaction proceeds according to a base-type mechanism.[133]

C. Synthesis of unstable intermediates. Spiro [2.5] octa-1,4-dien-3-one intermediate (II), which is easily solvated, cannot be obtained in the presence of homogeneous bases such as *t*-BuOK in *t*-BuOH, NaH in ether, or aqueous NaOH solution. It can, however, be obtained by passing a solution of (I) in ether through a column packed with KOH/Al_2O_3.[134)]

OH O⁻ O OH
Br⁻ CH_3OH
Br Br OCH_3
(I) (II)

Other acetylenic compounds which contain a group sensitive to alkali[135)] or a perchloro compound[136)] can also be obtained by using basic alumina catalysts. The solid base catalyst KOH/Al_2O_3 has a different selectivity for the hydrolysis of toluenesulfonate from that of the homogeneous base AcOK–AcOH,[137)] and is also more active in the hydrolysis of cholestan-3β-yl 3,5-dinitrobenzoate than a homogeneous base.[138)] Further, the yield of 7,7-dihalobicyclo [4.1.0] heptane (II, X=Cl or Br) from haloform and cylcohexene is found to be 12~15% with KOH/Al_2O_3,[136)] but less than 1% with KOH in aqueous solution.[139)]

$$X_3C\text{–}H \xrightarrow{KOH/Al_2O_3} X_3C^- \longrightarrow X^- + \underset{(I)}{X_2C:} \begin{cases} \xrightarrow{H_2O} CO + HCO_2^- \\ \xrightarrow{\text{cyclohexene}} \text{(II)} \end{cases}$$

This difference in yield is explained by taking into account the fact that water molecules retained in KOH/Al_2O_3 prepared by calcination react only with difficulty with carbenes (I), since the water molecules are tightly bound to the catalyst, whereas in an aqueous solution the free water molecules readily react with carbenes.

5.2.2 Correlation between catalytic activity and basic properties

A parallel has been found between the basic strengths of various solids as measured by the phenol vapour adsorption method and their catalytic activity for the dehydrogenation of isopropyl alcohol.[140] Malinowski *et al.* have investigated the reactions of formaldehyde with nitromethane,[141] acetaldehyde,[142] acetone,[143] and acetonitrile,[144] over silica gel catalysts containing various amounts of sodium at 275 °C. They found a linear relationship between the apparent reaction rate constants and the amount of sodium, as shown in Fig. 5–29.[145] Since

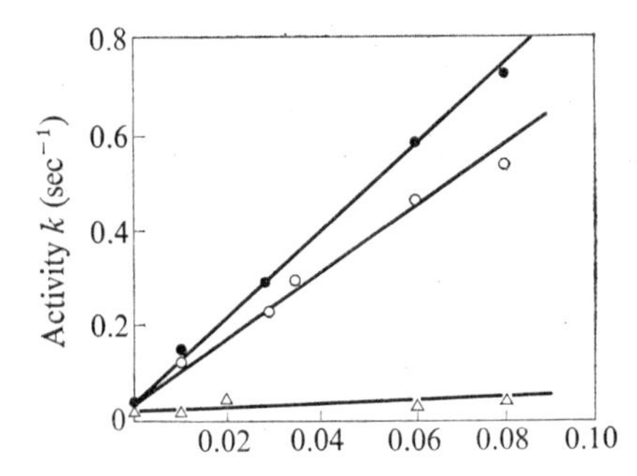

Sodium concentration (mol/100g SiO_2)

Fig. 5–29 Catalytic activities (reaction rate constants) *vs.* sodium content of silica gel catalysts
(●) $CH_3CHO + HCHO \rightarrow CH_2{=}CHCHO + H_2O$
(○) $CH_3COCH_3 + HCHO \rightarrow CH_2{=}CHCOCH_3 + H_2O$
(△) $CH_3CN + HCHO \rightarrow CH_2{=}CHCN + H_2O$

the basic strength of the catalyst is also directly proportional to the sodium concentration (*cf.* 3.1.1 and 3.2.1), this confirms the direct influence of basicity on reaction rates. On the other hand, a linear relationship also holds between the rate constants and the pK_a of the hydrogen donor molecules such as CH_3CHO, CH_3COCH_3 and CH_3CN.[145] Consequently, the reaction rate V is governed not only by the basic property X_B, but also by the acidic property K_A (the acid strength of the hydrogen donor molecule).

$$>C{=}O + H_2C< \xrightleftharpoons{\text{solid base}} >C{=}C< + H_2O$$

The relationship between V and X_B was examined in more detail for the reaction of formaldehyde with acetaldehyde by keeping K_A constant. As shown in Fig. 5–30, the rate constant increases linearly with increasing sodium content of the catalyst for the range below 0.81 mmol/g.[120] Since the reaction occurs even with a catalyst lacking

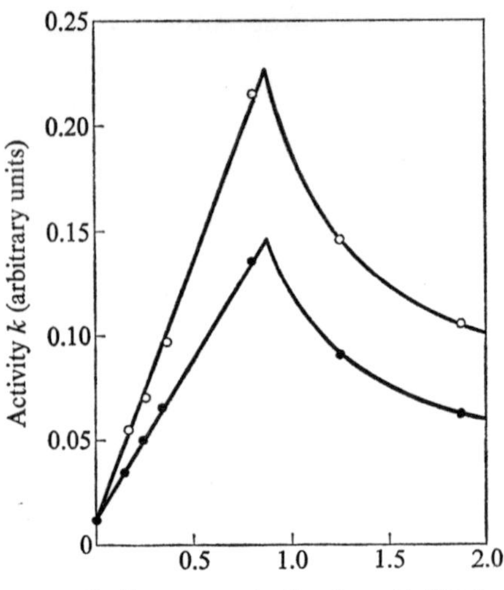

Fig. 5–30 Catalytic activities (reaction rate constants) *vs.* sodium concentration of silica gel catalysts
(●) reaction temperature: 275 °C
(○) reaction temperature: 300 °C

sodium, two types of active centre for this reaction are thought to exist: one would be the "sodium active –Si–O–Na site," and the other would include all those active sites which lack sodium. This is supported by the fact that the observed activation energy (10∼20 kcal/mol) for catalysts containing from 0.187 to 1.875 mmol/g Na differs from that for pure silica gel with a sodium/potassium content of 0.003 mmol/g (only 3 kcal/mol). The constancy of activation energy in the range of sodium content from 0.187∼1.875 mmol/g indicates that the reaction rate is proportional to the number of active centres containing sodium. In Fig. 5–30, the rate constant drops sharply with increasing sodium content in the range 1.25∼1.875 mmol/g. This is attributed to the decrease in the population of active sodium centres due to partial disruption of the silica gel structure by treatment with high concentrations of sodium hydroxide. The catalytic activity of MgO, CaO and PbO, heat-treated at various temperatures, in the aldol condensation of acetaldehyde, was also found to correlate well with the amount of base at $pK_a = +7.1$ and $+9.3$ measured by the indicator method (*cf.* 3.1.1).[121] A similar relationship between the catalytic activity of nitrous oxide-activated charcoal in the decomposition of hydrogen peroxide and the catalyst basicity as measured by the exchange method (*cf.* 3.2.2) has already been mentioned (see Table 3–3).[146]

Yamadaya *et al.* have measured the catalytic activities of alumina prepared by various means for the dehydration of alcohol, at the same time measuring the specific surface area and the basicity (by titration with bromothymol blue indicator).[147] As may be seen from the results shown in Fig. 5–31, the basicity increases at first with increasing surface

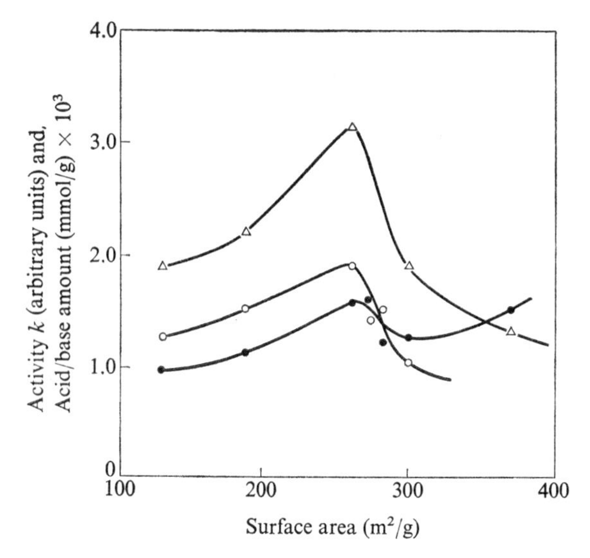

Fig. 5–31 Alcohol dehydration activity and amounts of acid and base on alumina catalysts *vs.* specific surface area
(△) catalytic activity, (○) amount of base
(●) amount of acid

area in the range 150~250 m^2/g, with a subsequent decrease at higher values. The change is approximately paralleled by that of the catalytic activity.

In the dehydrochlorination of 1,1,2-trichloroethane, $Cl_2HC(\alpha)$–$C(\beta)ClH_2$, catalyzed by basic metal oxides, two kinds of products are formed: 1,1-dichloroethylene (I), and 1,2-dichloroethylene (II).[148] Since (I) is formed by the abstraction of the more reactive α-proton, and (II) by the abstraction of the less reactive β-proton, it seems likely that stronger basic sites are required for the formation of (II), weaker sites being sufficient for the formation of (I). The assumption that the active sites for the formation of (I) and (II) are basic sites with $H_0 > +12.2$ and $H_0 > +18.4$ respectively enables calculation of the relative activities of SrO, CaO and MgO for the formation of (I) and (II) from the data for the distribution of basic strength given in Table 5–10. Calculation gives 1:1:6 and 1:7:20 respectively, whereas in fact the experimental relative activities were found to be 1:1.3:4.2 and 1:2.7:20.[148] Both sets of relative values seem to be in reasonable agreement. The selectivity of (I)/(II) is thus controlled by the basic strength of the catalysts. It is of interest here that the ratios of (I)/(II) and *trans*/*cis* are greater than unity in the case of the solid base catalyst, but less than unity for *trans*/*cis* and much less than unity for (I)/(II) for solid acid catalysts such as silica-alumina, alumina, alumina-boria, metal sulfates, etc.

The first order rate constant for the formation of benzyl benzoate from benzaldehyde over calcium oxides calcined at various temperatures is found to change in parallel with the change in catalyst basicity (see Fig. 3–2). Fig. 5–32 shows the good correlation between the catalytic activity and the amount of base per unit surface area.[149] In this reaction, the catalyst seems to be activated by adsorbing a very small amount of benzyl alcohol formed during the esterification reaction, so forming calcium benzylate. The long induction period normally observed in the reaction is considerably reduced by the addition of a small amount of benzyl alcohol or calcium benzylate. Thus, a calcium ion and an oxygen

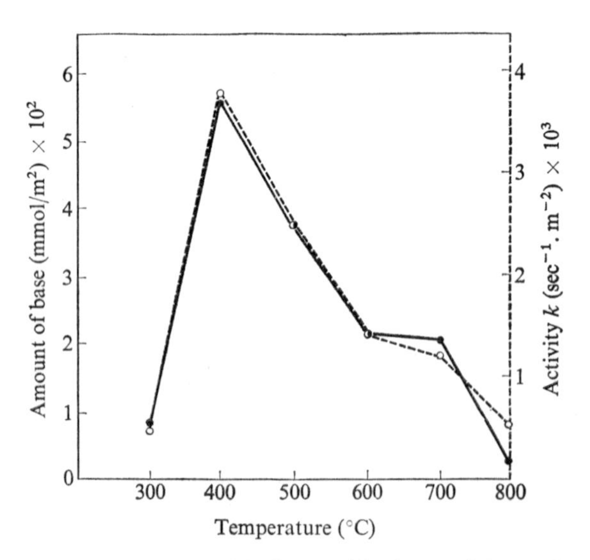

Fig. 5–32 Benzaldehyde esterification activity and amount of base *vs.* temperature of heat-treatment for $Ca(OH)_2$ catalysts
(—●—) amount of base,
(--○--) catalytic activity (rate constant k)

TABLE 5–10 Distribution of basic strength

Solid bases	Amount of base (mmol/g) at various ranges of H_0					
	+12.2~ +15.0	+15.0~ +17.2	+17.2~ +18.4	+18.4~ +26.5	$\geqq$ +26.5	Total
MgO†1	0.16	0.04	0.03	0.10	0	0.33
CaO†1	0.02	0	0	0.01	0.025	0.055
SrO†1	0	0	0.05	0	0.005	0.055
NaOH†2	0	0	0	1.75	0	1.75
Soda lime†1	0	0	0	0.43	0	0.43

†1 Evacuated at 450°C, 10^{-3} mmHg
†2 Evacuated at room temperature, 10^{-3} mmHg

ion on the catalyst surface are thought to act as a Lewis acid and a base for the adsorption of the alcohol as shown below.

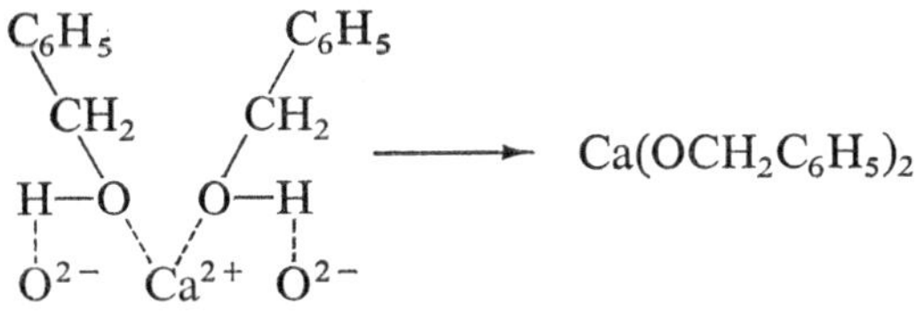

C_6H_5 C_6H_5
CH_2 CH_2
H—O O—H
O^{2-} Ca^{2+} O^{2-}

$$\longrightarrow Ca(OCH_2C_6H_5)_2$$

In the reaction of acrolein with ethanol on basic metal oxides, Krylov and Fokina have inferred from infrared studies that the alcoholate is first formed with the participation of metal and oxygen atoms on the solid bases.[133)]

5.3 Solid acid-base bifunctional catalysis

A catalyst on which two active groups, one acidic and one basic, are properly oriented is capable of powerful and specific catalysis even where the strength of the groups is only moderate. In the mutarotation of α-D-tetramethylglucose in benzene solution, Swain and Brown found that 2-hydroxypyridine is a powerful specific catalyst: at a concentration of 0.001 M it is 7×10^3 times as effective as a mixture of 0.001 M pyridine and 0.001 M phenol, though its basic strength is a factor of 10^{-4} less than that of pyridine, and its acidic strength 10^{-2} that of phenol.[150)] Since the velocity proves to be directly proportional to the concentration of 2-hydroxypyridine, it is clear that the catalytic mechanism is facilitated by the presence of an acidic and a basic group in the same catalyst molecule, see diagram below. This kind of mechanism

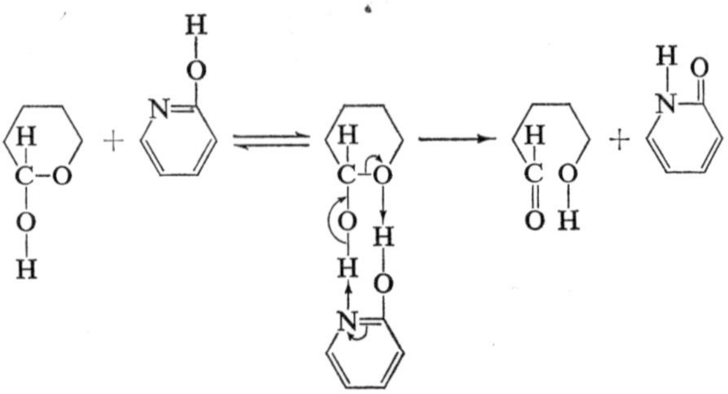

is known variously as the "concerted", "ternary", "push-pull", "synchronous" or "hydrogen-switch" type. Similarly, carboxylic acids are much more effective catalysts than phenols of comparable strength, probably because the carboxyl groups can act simultaneously as an acid and as a base,[151] i.e.,

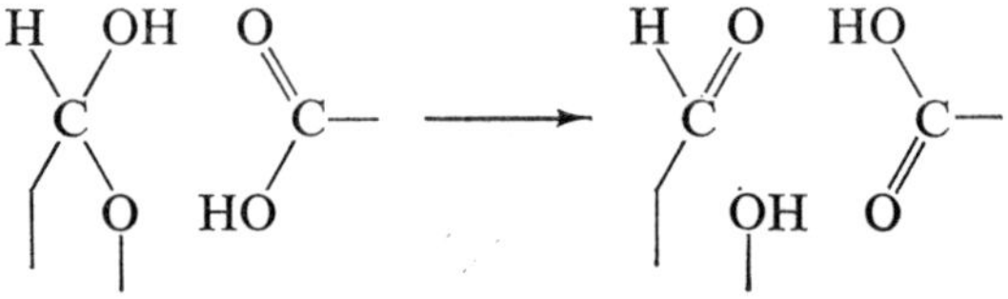

This kind of catalysis has been termed "acid-base bifunctional", and may be responsible for the action of enzymes. The activation of hydrogen by copper ions in aqueous solution is also considered to proceed by some kind of concerted mechanism, where cupric ions act like a Lewis acid, and hydrated water molecules as a base.[152]

$$H_2 + Cu(H_2O)_n^{2+} \longrightarrow \begin{matrix} H^- \text{-----} Cu^{2+} \\ H^+ \text{-----} H_2O \cdot (H_2O)_{n-1} \end{matrix}$$

$$\longrightarrow \begin{matrix} Cu^+H \\ H_3O^+ \cdot (H_2O)_{n-1} \end{matrix} \longrightarrow CuH(H_2O)_{n-1}^+ + H_3O^+$$

In heterogeneous reactions, Turkevich and Smith have shown that the isomerization of 1-butene to 2-butene is catalyzed by metal sulfates, sulfuric acid, phosphoric acid, etc.[153] (Note: the separations between acid and basic sites on sulfuric and phosphoric acid are 3.50 and 3.46 Å respectively.) Little activity was observed, however, for acetic acid, hydrogen chloride, etc. The higher catalytic activity of catalysts of the former group is attributed to acid-base bifunctional catalysis as illustrated in the diagram below. Horiuti independently advanced the same concept in interpreting the mechanism by which $Al_2O_3 \cdot ThO_2 \cdot H_2SO_4$ catalysts act in the synthesis of hydrogen cyanide from ammonia and carbon monoxide.[154] In the case of metal sulfates, bifunctional catalysis may be achieved by the vacant orbitals of the metal ions (induced Brønsted acid sites) acting as the acid, and the oxygen of the sulfate as the base, see diagram page 86. A similar mechanism has been suggested for the double-bond shift of olefins catalyzed by silica-alumina in order to explain the selectivity with respect to double-bond shift and *cis-trans* isomerization and the anomalously low *cis*/*trans* ratio of the 2-hexenes formed from *cis*-3-hexene.[61] The catalytic activity of alumina in alkene isomerization reactions is also attributable to the existence of acceptor-donor pairs on the surface.[155]

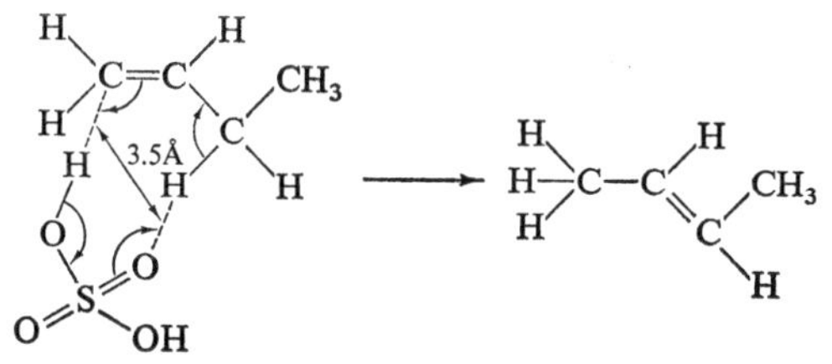

The relative values of the first-order rate constants mentioned in 5.1.1 for the depolymerization of paraldehyde catalyzed by $NiSO_4$, $CuSO_4$ and $SiO_2 \cdot Al_2O_3$ are (in terms of the unit acidity at an effective acid strength $H_0 \leqq -3$) 1,100:300:1. The difference may be attributed to the difference in the acid-base bifunctional catalytic modes of these catalysts. Misono and Yoneda associate the preferential formation of one isomer over the other in 1-butene isomerization catalyzed by a metal sulfate with the metal cation electronegativity.[156] Tanabe *et al.* have found that the mutarotation of α-D-tetramethylglucose in benzene, which is thought to involve acid-base bifunctional catalysis of the kind outlined above, is catalyzed by metal sulfates, silica-alumina, etc.[157]

The *trans-cis* isomerization of crotononitrile has been investigated using various catalysts including Al_2O_3, MgO, CaO, Na_2CO_3 and NaOH supported on silica gel, and some solid organic compounds.[158] Inorganic and organic compounds such as Al_2O_3, potassium 2-naphthol-3-carboxylate, sodium salicylate, etc., which have both acidic and basic groups, are found to be catalytically active. On the other hand, unmounted NaOH and Na_2CO_3, silica alone and potassium biphthalate, each of which possesses either only basic or only acidic properties, are inactive. These observations indicate that this isomerization too undergoes acid-base bifunctional catalysis.

The dehydration of various types of alcohols with alumina catalysts has been the object of intensive studies, and the reaction mechanism and the nature of the alumina have been discussed by Pines and Manassen.[159] They regard the dehydration of most alcohols (menthols, neomenthols, alkylcyclohexanols, decalols, bornanols, 2-phenyl-1-propanol, etc.) as taking the form of a *trans*-elimination, requiring the participation of both acidic and basic sites on the alumina. For this reason alumina may be thought of as a "solvating" agent, in that it must surround the alcohol molecule, thereby enabling the alumina acid sites to act as proton donors or electron acceptors, and the basic sites as the proton acceptors or electron donors.

As a typical example, the elimination scheme for *cis,cis*-1-decalol is shown in Fig. 5–33. The formation of 1,9- and *cis*-1,2-octalin clearly

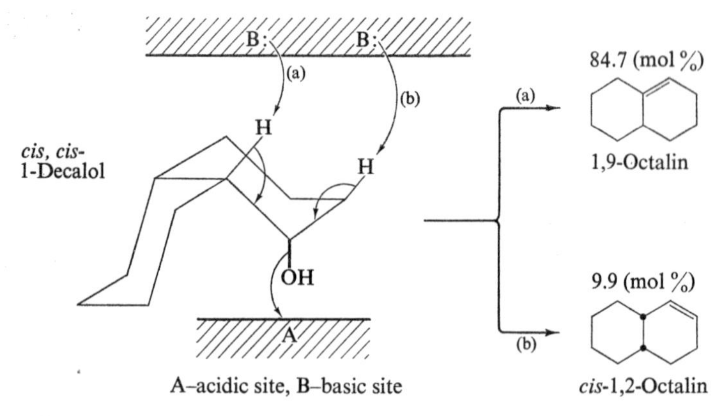

Fig. 5–33 Suggested scheme for *cis,cis*-1-decalol dehydration over Al_2O_3 catalysts

indicates that the dehydration undergoes a *trans*-elimination over alumina which acts both as a solvating agent and as an acid-base bifunctional catalyst.[156] In the dehydration of menthol over alumina catalysts, 2-menthene was found to be the predominant product.[159-60] 3-Menthene was also formed in each of the alcohol dehydrations, even when the extent of dehydration was small, as were traces of 1-menthene, although 1-menthene is not to be expected from a simple 1,2-elimination of the elements of water (see diagram p. 149). The preferential formation of 2-menthene is a clear indication of *trans*-elimination. This conclusion was further supported by the results obtained from the dehydra-

(I) (2-menthene) (II) (3-menthene) (III) (1-menthene)

	Composition (%)		
	(I)	(II)	(III)
(menthol) →	80 – 90	18 – 10	<2
(neomenthol) →	4 – 25	75 – 95	<1

tion of neomenthol, which yields 3-menthene as the predominant compound.

The elimination of hydrogen halide from alkyl halides in many cases is thought to proceed by a concerted mechanism with acid-base bifunctional catalysts similar to that for the alcohol water elimination reactions outlined above. In the elimination of HCl and HBr from 2, 3-dichlorobutane and 2,3-dibromobutane, KBO_2 and K_2CO_3 were found to show high stereoselectivity, with a large difference in the *cis/trans* ratio for the two forms of reactant (a concerted mechanism). $CaCl_2$, $CaSO_4$, $Ca_3(PO_4)_2$ showed no stereoselectivity and had equal *cis/trans* ratios (carbonium ion mechanism).[161] These results can be interpreted on the assumption that there is a two-fold interaction between catalyst and reactant (i.e. a two point adsorption), that is an interaction between cation acid and halogen (halide) as well as between anion base and hydrogen (proton). The stereoselectivity depends upon the relative intensity of these two interactions. The acceptor strength of the cation increases with increasing charge and decreasing size, while the donor strength of the anion may be related to the basicity of the ion in the liquid phase, or to the acidity of the corresponding acid. Although it is impossible to give figures for the basicity of anions in the crystal surface, it is likely that Cl^- is less basic than SO_4^{2-}, and SO_4^{2-} considerably less than PO_4^{3-}, CO_3^{2-} and BO_2^-. Should these two interactions be of comparable strength, the catalyst will be highly stereoselective. This must be the case with KBO_2, where a rather large monovalent cation (with a correspondingly low acceptor strength) is combined with a strongly basic anion.

Andréu *et al.* have studied further the eliminations of HBr from 1,1-diphenyl-2-bromoethylene and *cis*-bromostilbene and from 1-bromo-2-methyl-1-propene, *cis*- and *trans*-2-bromo-2-butene over $CaSO_4$, $Ca_3(PO_4)_2$, CaO and Li_2SO_4, and concluded that the presence of both electron acceptor and electron donor sites is necessary.[162] They also found that the catalysts with small cations and high electric charge together with anions of low basicity possess high activity but low selectivity. Those with larger cations and low electric charge and possessing strongly basic anions are more selective, but usually less active. However, it is not yet clear exactly how the activity and selectivity correlate with the acidic and/or basic properties of the catalyst surfaces described in Chapter 4.

Schwab and Kral have also suggested that, in a manner similar to the above elimination reactions, cumene cracking is facilitated by polarity within the surface of alumina between the normal acid centres and the basic centres.[163] The alumina catalyst can be improved by *n*-type doping (Table 5–11); in addition to the acidic sites, the basic sites

TABLE 5–11 Cumene cracking over doped alumina catalysts

Character	Dope	atom %	Activation energy q (kcal/mol)
—	—	0	54
p	Li_2O	2	57
p	Li_2O	6	62
n	GeO_2	3	48
n	V_2O_5	1	42
n	WO_3	0.25	41
n	WO_3	0.5	31
n	WO_3	1	26
n	WO_3	2	17.5
n	WO_3	2.5	17
n	WO_3	3	17.5
n	WO_3	4	22

in the form of mobile electrons and undissociated donor states created by the *n*-doping enhance the cracking activity.

The alumina sites which are active for the hydrogen-deuterium exchange in *n*-propane are considered by Pink *et al.* to be those weaker electron-transfer sites which, although capable of Lewis type acid-base reaction, are insufficiently strong to bring about the formation of radicals.[164] A possible mechanism for the exchange process may be formulated as follows:

$$\underset{\text{LA LB}}{\overline{\mathrm{C_3H_8}}} \longrightarrow \underset{\text{LA}\quad\text{LB}}{\overline{\mathrm{C_3H_7}\;\;\mathrm{H}}} \longrightarrow \underset{\text{LA}\quad\text{LB}}{\overline{\mathrm{C_3H_7}\;\;\mathrm{D}}} \longrightarrow \underset{\text{LA LB}}{\overline{\mathrm{C_3H_7D}}}$$

LA: Lewis acid, LB: Lewis base

Dissociative adsorption of propane is followed by hydrogen exchange at the site of the active hydrogen atom, with desorption of the deuterated species from the surface as the final step. This mechanism is in accord with the mass spectrometric evidence that the exchange of hydrogen in propane is a stepwise process involving replacement of not more than one hydrogen atom per adsorption-desorption cycle. The coordinated acid-base sites on alumina are possibly associated respectively with an abnormally exposed aluminium ion and a defect centre containing an excess of oxide ions.[165-6)] There is other evidence that the strengths of electron donor and acceptor sites may be critical in hydrogen-deuterium exchange. Tamaru *et al.* have shown that although hydrogen-deuterium exchange activity is found in a wide range of electron donor-acceptor (EDA) complexes, maximum activity is confined to a small range of complexes in which the electron affinity of the acceptor molecules lies at intermediate values.[167)] Complexes involving either stronger or weaker acceptor molecules are much less effective.

The heterogeneous catalysis of diethyl carbonate hydrolysis gives evidence of being an acid-base bifunctional catalysis, though the reaction does not proceed by a "concerted" mechanism of the type considered above. Sauer and Krieger showed that BeO, ZnO, PbO, NiO, NaCl/C (mounted on carbon), KCl/C, $CdSO_4$/C, CdO/C, etc. are active for the hydrolysis.[168)] On the basis of their kinetic results, which reveal that the reaction rate is independent of the water vapour concentration and proportional to the fraction of the surface covered with diethyl carbonate, and also that the values of activation energies and frequency factors are virtually the same for each of the catalysts, it seems that the ability of these catalysts to dissociate adsorbed water molecules adequately categorizes their activity. As shown in the diagram on p. 152, water molecules dissociated on the ionic surface of the catalysts act as Lewis base (or nucleophile) and Brønsted acid sites, although k_3 governs the over-all rate.

As described in Chapter 4, silica-alkaline earth metal oxides have both acidic and basic properties (*cf.* Table 4–8). In the case of these catalysts, experiments with poisoning by pyridine and phenol have suggested that while acid sites are active for the dehydration of *n*-

$$H_2O(g) \xrightarrow{k_1} \underset{\text{Catalyst}}{\underline{H^+\text{------}OH^-(ads.)}}$$

$$\underset{\text{Catalyst}}{\underline{H^+\text{------}OH^-(ads.)}} + RO\text{—}CO\text{—}OR \xrightarrow{k_2} \begin{array}{c} RO\text{—}CO\text{—}OR(ads.) \\ \underline{H^+\text{------}OH^-} \end{array}$$

$$\xrightarrow{k_3} \begin{array}{c} O^- \\ | \\ RO\text{—}\overset{\pm}{C}\text{—}OR \\ | \\ \underline{H^+\text{------}OH^-} \end{array} \xrightarrow{k_4} \begin{array}{c} \qquad\quad O \\ \qquad\quad \| \\ RO\text{------}C\text{—}OR \\ | \qquad\quad | \\ \underline{H \qquad\quad OH} \end{array}$$

$$\xrightarrow{k_5} HO\text{—}CO\text{—}OR(g) + ROH(g) \xrightarrow{k_6} ROH(g) + CO_2(g)$$

butanol, the basic sites are active in dehydrogenation.[169]

It will be of interest here to describe briefly concerted acid-base bifunctional catalysis of the kind exhibited by enzymes. A host of enzymatic transformations involve a generalized acid-plus-base mechanism.[170] There are many basic and acidic side-chains in proteins, and these are usually assumed to be in the immediate neighbourhood of the "active sites" X, Y, Z, at which the substrate molecule becomes bound. As illustrated in Fig. 5–34, the basic (B) and acidic (BH^+) groups could be substantially separated from the neighbourhood of the active sites

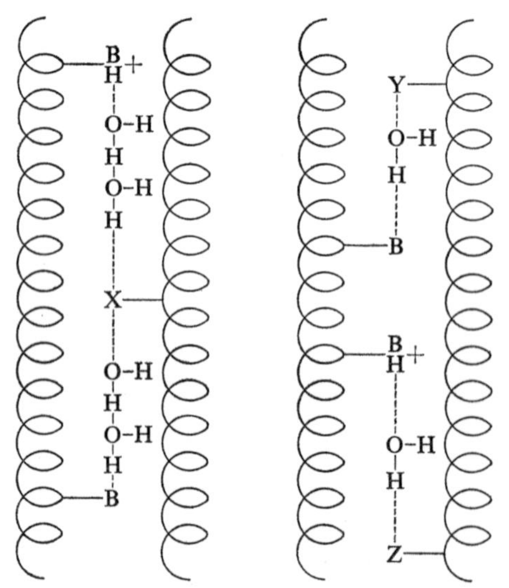

Fig. 5–34 Long range acid-base bifunctional catalysis
BH^+: acid site B: base site X, Y, Z: reaction points

and still perform their function of either donating or accepting a proton, through the medium of the chain of water molecules, hydrotactoids.[171] This picture offers a basis for interpreting the effects on enzymatic activity of a wide number of reagents which modify particular side-chains on the macromolecule, not all of which are likely to be in the vicinity of a site of substrate attachment. Modification of specific basic or acidic sites would, of course, directly remove their potential contribution to the catalysis, but so would changes in apolar side-chains, for these latter are instrumental in maintaining the water bridge between B or BH^+ and X, Y, or Z.[171] For a given solid acid-base bifunctional catalysis, provided that hydrotactoids can be formed by the creation of a polar surface between the acidic and basic sites, the proper orientation of both active sites would not necessarily be required for powerful and specific catalysis.

REFERENCES

1. M. W. Tamele, *Discussions Faraday Soc.*, **8,** 270 (1950).
2. O. Johnson, *J. Phys. Chem.*, **59,** 827 (1955).
3. K. Ikebe, N. Hara and K. Mita, *Kogyo Kagaku Zasshi*, **56,** 722 (1953).
4. R. L. Richardson and S. W. Benson, *J. Phys. Chem.*, **61,** 405 (1957).
5. G. A. Mills, E. R. Boedecker and A. G. Oblad, *J. Am. Chem. Soc.*, **72,** 1554 (1950).
6. T. H. Milliken, Jr., G. H. Mills and A. G. Oblad, *Discussions Faraday Soc.*, **8,** 279 (1950).
7. K. Tarama, S. Teranishi, H. Honda and S. Yoshida, *Shokubai (Tokyo)*, **6,** No. 4, 268 (1964).
8. J. B. Fisher and F. Sebba, *Actes Congr. Intern. Catalyse, 2^e^, Paris*, I, No. 29 (1960, Pub. 1961).
9. H. Uchida and H. Imai, *Shokubai (Tokyo)*, **3,** No. 2, 202 (1961).
10. T. Shirasaki, M. Okada, A. Kitahara, T. Mizushima, T. Okubo and K. Morikawa, *ibid.*, **5,** No. 3, 251 (1963).
11. M. Okada, T. Shirasaki, T. Mizushima and K. Morikawa, *ibid.*, **5,** No. 3, 255 (1963).
12. H. Pines and W. O. Haag, *J. Am. Chem. Soc.*, **82,** 2471 (1960).
13. E. Koberstein, *Z. Elektrochem.*, **64,** 909 (1960); L. Beranek, V. Bazant, L. M. Kraus and K. Kochloffl, *Collection Czech. Chem. Commun.*, **25,** 2513 (1960).
14. H. Pines and J. Ravoire, *J. Phys. Chem.*, **65,** 1859 (1961).
15. S. E. Tung and E. Meininch, *J. Catalysis*, **3,** 229 (1964).
16. J. B. Peri, *J. Phys. Chem.*, **69,** 231 (1965).
17. E. Echigoya, *Nippon Kagaku Zasshi*, **76,** 1142 (1955).
18. J. Kiji, K. Hirota and Y. Kobayashi, *Shokubai (Tokyo)*, **3,** No. 2, 198 (1961).
19. E. N. Rosolovskaja, O. L. Shakhnovskaja and K. V. Topchieva, *Kinetika i Kataliz*, **7,** 750 (1966).
20. T. Nishizawa, H. Hattori, T. Uematsu and T. Shiba, *Intern. Congr. Catalysis, 4th, Moscow, Preprints of Papers*, No. 55 (1968); T. Shiba, *Japan–U.S.A.*

Seminar Catalytic Sci., Tokyo and Kyoto, Preprints of Papers from Japan, No. J–6–1 (1968).
21. Y. Matsumoto and Y. Morita, *Nenryo Kyokaishi*, **46,** 168 (1967).
22. Y. Murakami, H. Nozaki and J. Turkevich, *Shokubai* (*Tokyo*), **5,** No. 3, 262 (1963).
23. Y. Izumi and T. Shiba, *Bull. Chem. Soc. Japan*, **37,** 1797 (1964).
24. Y. Watanabe, *Dai 18-nenkai Kōenyokō-shū* (Japanese) (Ann. Meeting Chem. Soc. Japan, 18th, Tokyo, Preprints of Papers), No. 2330 (1965).
25. V. A. Dzisko, *Proc. Intern. Congr. Catalysis, 3rd, Amsterdam*, I, No. 19 (1964).
26. N. Pernicone, G. Liberti and L. Ersini, *Intern. Congr. Catalysis, 4th, Moscow, Preprints of Papers*, No. 21 (1968).
27. M. Tani and Y. Ogino, *Shokubai* (*Tokyo*), **8,** No. 3, 250 (1966).
28. K. Tanabe and R. Ohnishi, *J. Res. Inst. Catalysis, Hokkaido Univ.*, **10,** 229 (1962).
29. K. Tanabe and C. Mugiya, *ibid.*, **14,** 101 (1966).
30. K. Tanabe and T. Takeshita, *Advances in Catalysis*, vol. 17, p. 315, Academic Press, 1967.
31. K. Tanabe and A. Aramata, *J. Res. Inst. Catalysis, Hokkaido Univ.*, **8,** 43 (1960).
32. R. P. Bell, O. M. Lidwell and M. M. Vaughan-Jackson, *J. Chem. Soc.*, **1936,** 1792.
33. K. Tanabe and Y. Watanabe, *J. Res. Inst. Catalysis, Hokkaido Univ.*, **7,** 120 (1959).
34. K. Tanabe, *Kagaku To Kogyo* (*Tokyo*), **19,** 1180 (1966).
35. H. Takida and K. Noro, *Kobunshi Kagaku*, **21,** 23, 109 (1964).
36. K. Tarama, S. Teranishi, K. Hattori and T. Ishibashi, *Shokubai* (*Tokyo*), **4,** No. 1, 69 (1962).
37. Y. Watanabe and K. Tanabe, *J. Res. Inst. Catalysis, Hokkaido Univ.*, **12,** 56 (1964).
38. H. A. Benesi, *J. Phys. Chem.*, **61,** 970 (1957).
39. U.S. Pat. 1,994,035 (1935); Brit. Pat. 628,165 (1946); U.S. Pat. 2,426,638 (1947); U.S. Pat. 2,426,637 (1947); U.S. Pat. 2,478,201 (1949); U.S. Pat. 2,478,932 (1949); U.S. Pat. 2,598,411 (1952); U.S. Pat. 2,637,748 (1953); U.S. Pat. 2,691,053 (1954); U.S. Pat. 2,767,227 (1956); Brit. Pat. 800,758 (1958).
40. S. Okazaki, *Shokubai* (*Tokyo*), **10,** No. 4, 242 (1968).
41. Y. Ogino, *ibid.*, **4,** No. 1, 73 (1962).
42. T. Kawakami and Y. Ogino, *ibid.*, **10,** (23rd Symp. Catalysis, Preprints of Papers), 136 (1968).
43. T. Kagiya, T. Sano, T. Shimidzu and K. Fukui, *Kogyo Kagaku Zasshi*, **66,** 1893 (1963).
44. K. Kearby, *Actes Congr. Intern. Catalyse, 2e, Paris*, III, No. 134, (1960, Pub. 1961).
45. A. Tada, Y. Yamamoto, M. Ito and A. Suzuki, *Dai 22-nenkai Kōenyokō-shū* (Japanese) (Ann. Meeting Chem. Soc. Japan, 22nd, Tokyo, Preprints of Papers), No. 06421 (1969).
46. K. Tanaka and A. Ozaki, *J. Catalysis*, **8,** 1 (1967); *Bull. Chem. Soc. Japan*, **41,** 2812 (1968).
47. K. Tanaka and A. Ozaki, *Dai 22-nenkai Kōenyokō-shū* (Japanese) (Ann. Meeting Chem. Soc. Japan, 22nd, Tokyo, Preprints of Papers), No. 06406 (1969).

48. H. Ohtsuka and K. Aomura, *Bull. Japan Petrol. Inst.*, **4,** 3 (1962); *Yuki Gosei Kagaku Kyokai Shi*, **21,** 737 (1963); K. Aomura, *Kagaku Kojo*, **5,** 47 (1961).
49. T. Yamanaka, *Shokubai Kagaku* (Japanese), p. 211, Nikkan Kōgyō Shinbun Sha, 1956.
50. A. Mitsutani and Y. Hamamoto, *Kogyo Kagaku Zasshi*, **67,** 1231 (1964).
51. Y. Hamamoto and A. Mitsutani, *ibid.*, **67,** 1227 (1964).
52. M. Misono and Y. Yoneda, *Bull. Chem. Soc. Japan*, **40,** 42 (1967).
53. Y. Yoneda, *J. Catalysis*, **9,** 51 (1967).
54. Y. Yoneda, *Dai 5-kai Hannō Kōgaku Shinpojium Kōenyōshi-shū* (Japanese), No. 14, Soc. Chem. Engr., Japan (1965).
55. W. H. Wilmot, R. T. Barth and D. S. MacIver, *Proc. Intern. Congr. Catalysis, 3rd, Amsterdam*, II, No. 6 (1964).
56. M. Misono, Y. Saito and Y. Yoneda, *ibid.*, I, No. 18 (1964).
57. V. A. Dzisko, M. Kolovertnova, T. S. Vinnikova and Yu. O. Bulgakova, *Kinetika i Kataliz*, **7,** 655 (1966).
58. K. Hirota, T. Imanaka and Y. Kobayashi, *Shokubai* (*Tokyo*), **3,** No. 2, 206 (1961).
59. A. Clark, V. C. F. Holm and D. M. Blackburn, *J. Catalysis*, **1,** 244 (1962); A. Clark and V. C. F. Holm, *ibid.*, **2,** 16, 21 (1963).
60. P. L. Hsieh, *ibid.*, **2,** 211 (1963).
61. D. M. Brouwer, *ibid.*, **1,** 22 (1962).
62. D. S. MacIver, W. H. Wilmot and J. M. Bridges, *ibid.*, **3,** 502 (1964).
63. C. J. Norton, *Ind. Eng. Chem. Process Design Develop.*, **3,** 231 (1964).
64. M. Sato, T. Aonuma and T. Shiba, *Proc. Intern. Congr. Catalysis, 3rd, Amsterdam*, I, No. 17 (1964).
65. T. Yamaguchi and K. Tanabe, *Intern. Congr. Catalysis, 4th, Moscow, Preprints of Papers*, No. 80 (1968).
66. R. Ohnishi, T. Takeshita and K. Tanabe, *Shokubai* (*Tokyo*), **7,** No. 3, 306 (1965).
67. K. Tanabe and T. Takeshita, *San Enki Shokubai* (Japanese), p. 198, Sangyō Tosho, 1965.
68. F. E. Shephard, J. J. Rooney and C. Kemball, *J. Catalysis*, **1,** 379 (1962).
69. V. C. F. Holm, G. C. Bailey and A. Clark, *J. Phys. Chem.*, **63,** 129 (1959).
70. J. J. Rooney and R. C. Pink, *Trans. Faraday Soc.*, **58,** 1632 (1962).
71. D. M. Brouwer, *J. Catalysis*, **1,** 372 (1962); *Chem. Ind.* (*London*), 177 (1961).
72. W. I. Aalbersberg, J. Gaaf and E. L. Mackor, *J. Chem. Soc.*, **1961,** 905.
73. A. E. Hirschler and J. O. Hudson, *J. Catalysis*, **3,** 239 (1964).
74. T. Shiba, T. Aonuma, K. Yoshida, H. Hattori and M. Sato, *Shokubai* (*Tokyo*), **6,** No. 3, 176 (1964).
75. J. W. Ward and R. C. Hansford, *J. Catalysis*, **13,** 154 (1969).
76. I. D. Chapman and M. L. Hair, *ibid.*, **2,** 145 (1963).
77. K. Tarama, S. Teranishi, S. Yoshida, H. Honda and S. Taniguchi, *Dai 18-nenkai Kōenyokō-shū* (Japanese) (Ann. Meeting Chem. Soc. Japan, 18th, Osaka, Preprints of Papers), No. 2409 (1965).
78. T. V. Antipina, O. V. Bulgakov and A. V. Uvarov, *Intern. Congr. Catalysis, 4th, Moscow, Preprints of Papers*, No. 77 (1968).
79. J. B. Peri, *Actes Congr. Intern. Catalyse, 2^e^, Paris*, II, No. 64 (1960, Pub. 1961).
80. A. Ozaki and K. Kimura, *J. Catalysis*, **3,** 395 (1964).
81. J. W. Ward, *ibid.*, **9,** 225 (1967); **10,** 34 (1968); **11,** 238, 251 (1968).
82. K. V. Topchieva, B. V. Romanovsky, L. I. Piguzova, Ho si Thoang and Y.

W. Bizreh, *Intern. Congr. Catalysis, 4th, Moscow, Preprints of Papers*, No. 57 (1968).
83. H. A. Benesi, *J. Catalysis*, **8,** 368 (1967).
84. T. Aonuma, M. Sato and T. Shiba, *Shokubai* (*Tokyo*), **5,** No. 3, 274 (1963).
85. H. Bremer and K. H. Steinberg, *Intern. Congr. Catalysis, 4th, Moscow, Preprints of Papers*, No. 76 (1968).
86. R. P. Bell and B. G. Skinner, *J. Chem. Soc.*, **1952,** 2955.
87. K. Tanabe and M. Matsuda, *J. Res. Inst. Catalysis, Hokkaido Univ.*, **9,** 246, 239 (1961).
88. N. Kominami and C. Sakurai, *Dai 17-nenkai Kōenyokō-shū* (Japanese) (Ann. Meeting Chem. Soc. Japan, 17th, Tokyo, Preprints of Papers), No. 2V18 (1964).
89. K. Tanabe, I. Kashiki, T. Takeshita and R. Ohnishi, unpublished data.
90. I. Matsuzaki, Y. Fukuda and K. Tanabe, unpublished data.
91. Neth. Pat Appl. 6,506,830 (1965); Japan. Pat. Shōwa 42–6,894 (1967).
92. S. Enomoto and M. Inowe, *Shōwa 43-nendo Shokubai Kenkyū Happyōkai Kōenyōshi-shū* (Japanese) (Meeting Catalysis Soc. Japan, Sendai, Abstr. Papers), No. 2 (1968).
93. T. Kotanigawa, *Shūki Taikai Kōenyokō-shū* (Japanese), No. 3E–04, Chem. Soc. of Japan (1969).
94. H. R. Gerberich and W. K. Hall, *J. Catalysis*, **5,** 99 (1966).
95. L. de Mourgues, D. Barthomeuf, F. Figueras, M. Perrin, Y. Trambouze and M. Prettre, *Intern. Congr. Catalysis, 4th, Moscow, Preprints of Papers*, No. 61 (1968).
96. M. Nitta, I. Isa, I. Matsuzaki and K. Tanabe, *Dai 21-nenkai Kōenyōkō-shū* (Japanese) (Ann. Meeting Chem. Soc. Japan, 21st, Osaka, Preprints of Papers), No. 18206 (1968).
97. T. Yamaguchi, K. Matsuda and K. Tanabe, *Dai 22-nenkai Kōenyōkō-shū* (Japanese) (Ann. Meeting Chem. Soc. Japan, 22nd, Tokyo, Preprints of Papers), No. 06407 (1969).
98. T. Takeshita, K. Arata, T. Sano and K. Tanabe, *Kogyo Kagaku Zasshi*, **69,** 916 (1966).
99. L. M. Stock and H. C. Brown, *Advances in Physical Organic Chemistry*, vol. 1, p. 35, Academic Press, 1963.
100. G. A. Olah, S. J. Kuhn and S. H. Flood, *J. Am. Chem. Soc.*, **84,** 1688 (1962).
101. Japan. Pat. Shōwa 38–10,997 (1963), 39–9,696 (1964), 39–18,963 (1964); Brit. Pat. 950,782 (1964) [*CA*, **60,** 13348f (1964)].
102. Japan. Pat. Shōwa 35–6,897 (1960), 35–5,997 (1960), 35–9,646 (1960).
103. O. V. Krylov, *Probl. Kinetiki i Kataliza, Acad. Nauk, USSR*, vol. 13, p. 151, 1968.
104. F. N. Hill, F. E. Bailey and J. T. Fitzpatrick, *Ind. Eng. Chem.*, **50,** 5 (1958).
105. O. V. Krylov, M. J. Kushnerev, Z. A. Markova and E. A. Fokina, *Proc. Intern. Congr. Catalysis, 3rd, Amsterdam*, I, No. 74 (1964).
106. Japan. Pat. Shōwa 18–1048 (1943).
107. T. Kagiya, T. Sano and K. Fukui, *Kogyo Kagaku Zasshi*, **67,** 951 (1964).
108. R. N. Goyce, *J. Polymer Sci.*, **3,** 169 (1948).
109. N. Ogata and S. Yumoto, *Bull. Chem. Soc. Japan*, **31,** 907 (1958).
110. Brit. Pat. 958,161 (1964) [*CA*, **61,** 5434c (1964)].
111. France Pat. 1,356,267 (1964) [*CA*, **61,** 4211g (1964)].
112. U.S. Pat. 3,128,318 (1964) [*CA*, **61,** 1754b (1964)].

113. H. Pines and L. A. Shaap, *Advances in Catalysis*, vol. 12, p. 120, Academic Press, 1960.
114. W. O. Haag and H. Pines, *J. Am. Chem. Soc.*, **82**, 387 (1960).
115. Y. Schächter and H. Pines, *J. Catalysis*, **11**, 147 (1968).
116. A. Clark and J. N. Finch, *Intern. Congr. Catalysis, 4th, Moscow, Preprints of Papers*, No. 75 (1968).
117. N. F. Foster and R. J. Cvetanović, *J. Am. Chem. Soc.*, **82**, 4274 (1960).
118. M. Albeck, E. Gil-Av, Ch. Rav-Acha and Y. Schächter, *Israel J. Chem.*, **5**, 76 (1967).
119. F. M. Scheidt, *J. Catalysis*, **3**, 372 (1964).
120. S. Malinowski and S. Basinski, *ibid.*, **2**, 203 (1963).
121. T. Kawaguchi, S. Hasegawa, S. Morikawa and H. Suzuki, *Shōwa 43-nendo Shokubai Kenkyū Happyōkai Kōenyōshi-shū* (Japanese) (Meeting Catalysis Soc. Japan, Sendai, Abstr. Papers), No. 3 (1968).
122. K. Fukui and M. Takei, *Bull. Inst. Chem. Res., Kyoto Univ.*, **26**, 85 (1951); *CA*, **49**, 5426e (1955).
123. T. Ishikawa and T. Kamio, *Tokyo Kogyo Shikensho Hokoku*, **58**, 40 (1963).
124. M. J. Astle and W. C. Gergel, *J. Org. Chem.*, **21**, 493 (1956).
125. E. D. Bergmann and R. Corett, *ibid.*, **23**, 1507 (1958); **21**, 167 (1956).
126. M. J. Astle and R. W. Etherington, *Ind. Eng. Chem.*, **44**, 2871 (1952).
127. C. J. Schmidle and R. C. Mansfield, *ibid.*, **44**, 1388 (1952).
128. M. J. Astle and F. P. Abatt, *J. Org. Chem.*, **21**, 1228 (1956).
129. France Pat. 1,273,409 (1962) [*CA*, **57**, 9743d (1962)].
130. I. Mochida, J. Take, Y. Saito and Y. Yoneda, *Shokubai* (*Tokyo*), **8**, No. 3, 242 (1966).
131. P. Andréu, M. Heunisch, E. Schmitz and H. Noller, *Z. Naturforsch.*, **196**, 649 (1964).
132. H. Noller, H. Hantsche and P. Andréu, *J. Catalysis*, **4**, 354 (1965).
133. O. V. Krylov and E. A. Fokina, *Intern. Congr. Catalysis, 4th, Moscow, Preprints of Papers*, No. 64 (1968).
134. S. Winstein and R. Baird, *J. Am. Chem. Soc.*, **79**, 4238 (1957); **85**, 567 (1963).
135. G. Belil, J. Castellá, J. Castells, R. Mestres, J. Pascual and F. Serratosa, *Anales Real Soc. Espan. Fis. Quim.* (*Madrid*), *Ser. B*, **57**, 614 (1961).
136. F. Serratosa, *J. Chem. Educ.*, **41**, 564 (1964).
137. J. Castells and G. A. Fletcher, *J. Chem. Soc.*, **1956**, 3245.
138. G. H. Douglas, P. S. Ellington, G. D. Meakins and R. Swindells, *ibid.*, **1959**, 1720.
139. W. von E. Doering and A. K. Hoffmann, *J. Am. Chem. Soc.*, **76**, 6162 (1954).
140. O. V. Krylov and E. A. Fokina, *Probl. Kinetiki i Kataliza, Acad. Nauk, USSR*, vol. 8, p. 248, 1955.
141. S. Malinowski, H. Jedrzejewski, S. Basinski, Z. Lipinski and J. Moszezenska, *Roczniki Chem.*, **30**, 1129 (1956).
142. S. Malinowski, S. Basinski, M. Olszewska and H. Zieleniewska, *ibid.*, **31**, 123 (1957).
143. S. Malinowski, W. Kiewlicz and E. Soltys, *Bull. Soc. Chim.*, 439 (1963).
144. S. Malinowski, S. Benbenek, I. Pasynktewicz and E. Wojciechowska, *Roczniki Chem.*, **32**, 1089 (1958).
145. S. Malinowski, S. Basinski, S. Szczepanska and W. Kiewlicz, *Proc. Intern. Congr. Catalysis, 3rd, Amsterdam*, I, No. 20 (1964).
146. E. Naruko, *Kogyo Kagaku Zasshi*, **67**, 2019 (1964).

147. M. Yamadaya, K. Shimomura, T. Konoshita and H. Uchida, *Shokubai* (*Tokyo*), **7,** No. 3, 313 (1965).
148. J. Take, N. Kikuchi and Y. Yoneda, *ibid.*, **10,** (23rd Symp. Catalysis, Preprints of Papers), 127 (1968).
149. K. Saito and K. Tanabe, *ibid.*, **11,** No. 4, 206P (1969).
150. C. G. Swain and J. F. Brown, Jr., *J. Am. Chem. Soc.*, **74,** 2534, 2538 (1952).
151. A. M. Eastham, E. L. Blackall and G. A. Latremouille, *ibid.*, **77,** 2182 (1955); E. L. Blackall and A. M. Eastham, *ibid.*, **77,** 2184 (1955).
152. J. Halpern, *Advances in Catalysis*, vol. 11, p. 301, Academic Press, 1959.
153. J. Turkevich and R. K. Smith, *J. Chem. Phys.*, **16,** 446 (1948).
154. J. Horiuti, *Shokubai* (*Sapporo*), **1,** 67 (1948).
155. L. V. Nicolescu, A. Nicolescu, M. Gruia, A. Terlecki-Baricevic, M. Dardan, Em. Angelescu and A. Ionescu, *Intern. Congr. Catalysis, 4th, Moscow, Preprints of Papers*, No. 78 (1968).
156. M. Misono and Y. Yoneda, *Dai 19-nenkai Kōenyokō-shū* (Japanese) (Ann. Meeting Chem. Soc. Japan, 19th, Yokohama, Preprints of Papers), No. 12A-123 (1966).
157. K. Tanabe, A. Nagata and T. Takeshita, *J. Res. Inst. Catalysis, Hokkaido Univ.*, **15,** 181 (1967).
158. K. Ohki, K. Fujita, A. Miyamura and A. Ozaki, *Shokubai* (*Tokyo*), **7,** No. 3, 309 (1965).
159. H. Pines and J. Manassen, *Advances in Catalysis*, vol. 16, p. 49, Academic Press, 1966.
160. H. Pines and C. N. Pillai, *J. Am. Chem. Soc.*, **83,** 3270 (1961).
161. H. Noller, P. Andréu, E. Schmitz, S. Serain, O. Neufang and J. Giron, *Intern. Congr. Catalysis, 4th, Moscow, Preprints of Papers*, No. 81 (1968).
162. P. Andréu, H. Noller, S. Perozo, E. Schmitz and P. Tovar, *ibid.*, No. 82 (1968).
163. G.-M. Schwab and H. Kral, *Proc. Intern. Congr. Catalysis, 3rd, Amsterdam*, I, No. 20 (1964).
164. B. D. Flockhart, S. S. Uppal, I. R. Leith and R. C. Pink, *Intern. Congr. Catalysis, 4th, Moscow, Preprints of Papers*, No. 79 (1968).
165. B. D. Flockhart, J. A. N. Scott and R. C. Pink, *Trans. Faraday Soc.*, **62,** 730 (1966).
166. J. B. Peri, *J. Phys. Chem.*, **69,** 220 (1965).
167. M. Ichikawa, M. Soma, T. Ohnishi and K. Tamaru, *Trans. Faraday Soc.*, **63,** 2012 (1967).
168. R. W. Sauer and K. A. Krieger, *J. Am. Chem. Soc.*, **74,** 3116 (1952).
169. H. Niiyama and E. Echigoya, *Shokubai* (*Tokyo*), **10,** (23rd Symp. Catalysis, Preprints of Papers), 129 (1968).
170. M. Bender, *Chem. Rev.*, **60,** 53 (1960).
171. I. M. Klotz, *Horizons in Biochemistry* (ed. M. Kasha *et al.*), p. 523, Academic Press, 1962.

Chapter 6

Conclusion and Future Problems

Various metal oxides, sulfides, sulfates, phosphates, carbonates, hydroxides and halides, mixed metal oxides, natural clay minerals, mounted acids and bases and activated carbons have been shown to possess acidic and/or basic properties, which are markedly dependent upon the method of preparation, heat-treatment, and susceptible to poisoning, etc. The several structures which have been proposed for various acidic and basic centres have also been reviewed.

The characteristics and special features of solid acid and base catalysts in comparison with homogeneous catalysts have been emphasized on the basis of correlations between their acidic and/or basic properties and their catalytic activity and selectivity, and data derived from other kinetic studies. In a few cases the reaction mechanism has also been discussed.

A number of solid acids and bases which have been discovered recently are expected to find applications as catalysts with novel activity and selectivity in petrochemical reactions, and as catalysts effective in organic synthesis. It is possible that many more new solid acids and bases will be discovered; combinations of the various metal oxides offer particularly promising possibilities.

As will have been clear from the discussion in 4.2, the mixed oxides which are well-known as solid acids or bases are limited to a very few species ($SiO_2 \cdot Al_2O_3$, $SiO_2 \cdot MgO$, $SiO_2 \cdot ZrO_2$, $Al_2O_3 \cdot B_2O_3$, $Al_2O_3 \cdot MgO$), each of which contains either SiO_2 or Al_2O_3, both essential constituents of the clay minerals. These combinations of oxides are shown by the half-tone lines in Table 6–1. Other combinations, for which only the acidic properties have been measured, are shown by solid lines. A few, e.g. $TiO_2 \cdot ZnO$, do not contain any clay mineral components. Many other combinations within the region of the periodic table indicated are expected to exhibit acidic and/or basic properties. It is to be hoped that these properties will be detected and measured in the near future, to be duly correlated with the various factors which studies to date have shown relevant; the electronegativity, charge, ionic radius, etc. of the oxide metal ion, the distance between the metal and oxygen atoms, and (of course) the catalytic activity and selectivity. The eventual aim should be, if possible, to tabulate the surface properties of mixed

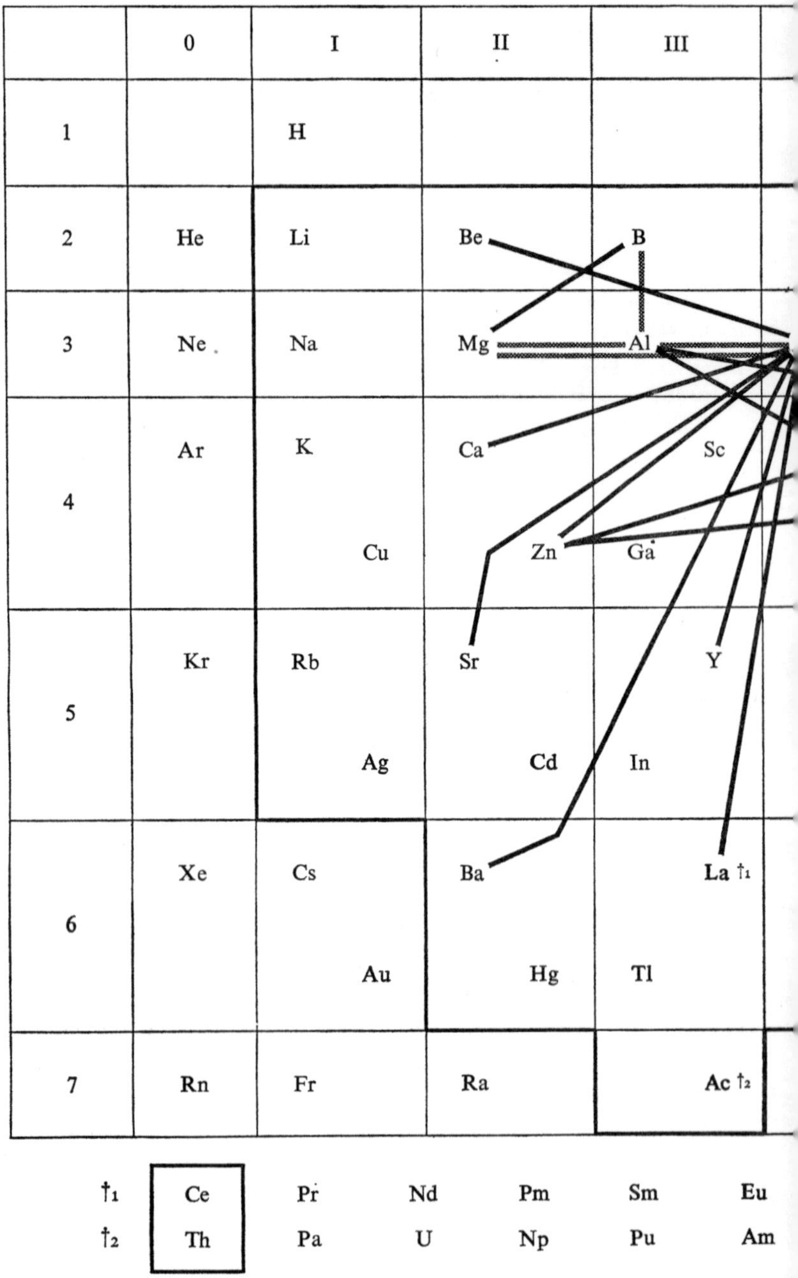

	0	I	II	III
1		H		
2	He	Li	Be	B
3	Ne	Na	Mg	Al
4	Ar	K Cu	Ca Zn	Sc Ga
5	Kr	Rb Ag	Sr Cd	Y In
6	Xe	Cs Au	Ba Hg	La †1 Tl
7	Rn	Fr	Ra	Ac †2

†1	Ce	Pr	Nd	Pm	Sm	Eu
†2	Th	Pa	U	Np	Pu	Am

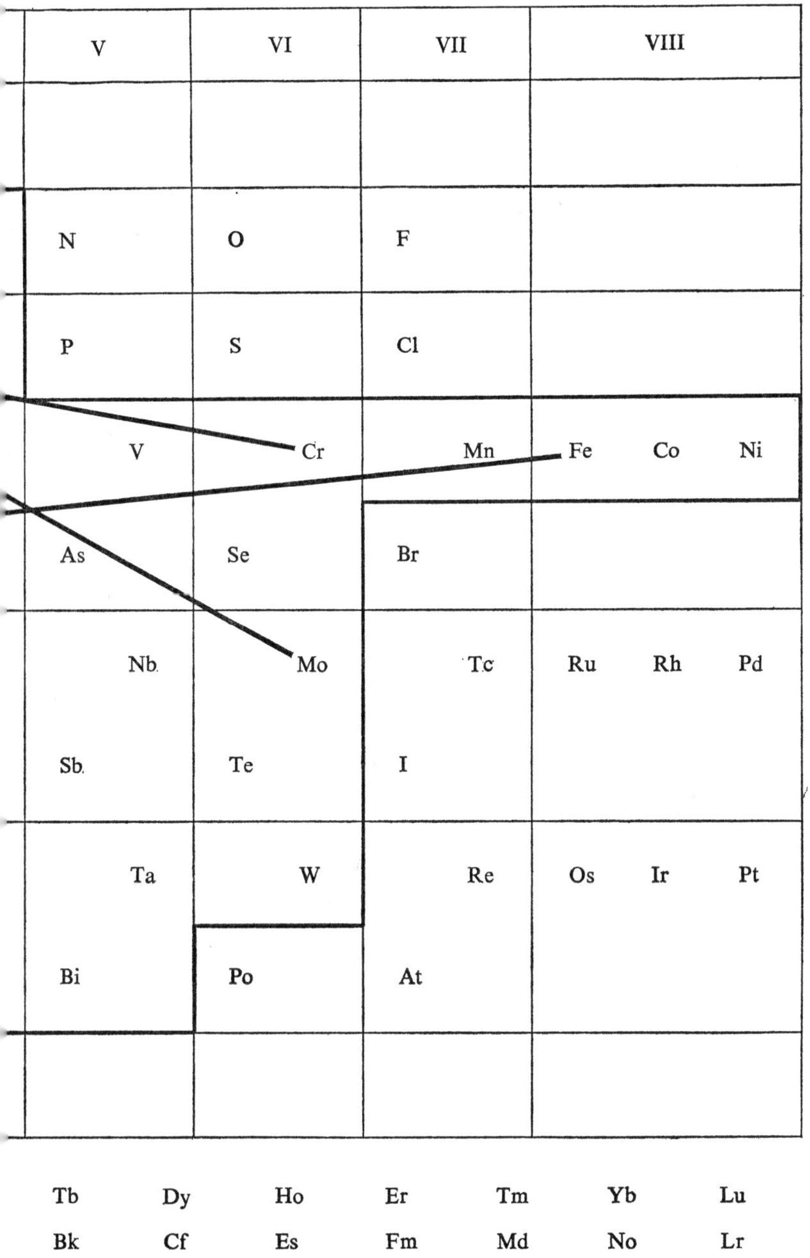

V	VI	VII	VIII		
N	O	F			
P	S	Cl			
V	Cr	Mn	Fe	Co	Ni
As	Se	Br			
Nb	Mo	Tc	Ru	Rh	Pd
Sb	Te	I			
Ta	W	Re	Os	Ir	Pt
Bi	Po	At			

Tb	Dy	Ho	Er	Tm	Yb	Lu
Bk	Cf	Es	Fm	Md	No	Lr

oxide acids and bases in the form of a kind of "sub-table" of the periodic table. The establishment of such a table would inevitably have to include full reference to the effects of not only changes in the composition of the mixed oxides, but also their methods of preparation and conditions of heat-treatment.

All studies of solid acids and bases depend ultimately on accurate measurements of the surface acid and basic properties. Improvements of technique here are vitally important for further advances in this field of research. There is a particularly urgent need for more quantitative methods with wider application to the measurement of acid-base properties on the surface of coloured solid materials.

Future investigations must, of necessity, grapple increasingly with the problems of the structure of acid and/or basic sites, their catalytic action, and the details of the reaction mechanisms involved.

Author Index

Numbers in parentheses are reference numbers and indicate that an author's work is referred to although his name may not be cited in the text. Numbers in italics show the page on which the complete reference is listed.

A

Aalbersberg, W. I., 127 (72), *155*
Abatt, F. P., 139 (128), *157*
Adams, C. J., 12 (24), 22 (24), *31*
Albeck, M., 138 (118), *157*
Amenomiya, Y., 22 (50), *31*
Amphlett, C. B., 85 (119), *100*
Andréu, P., 139 (131, 132), *157*, 149 (161), 150 (162), *158*
Angelescu, Em., 147 (155), *158*
Angell, C. L., 78 (102), 80 (108), *100*
Antipina, T. V., 48 (12), *97*, 129 (78), *155*
Aomura, K., 118 (48), *155*
Aonuma, T., 13 (27), *31*, 124 (64), 126 (64), 127 (64, 74), *155*, 131 (84), *156*
Arai, H., 26 (79), *32*
Arai, Y., 74 (90), 75 (90), *100*
Aramata, A., 109 (31), *154*
Arata, K., 135 (98), *156*
Arrhenius, S. A., 1 (1), *3*
Astle, M. J., 139 (124), 139 (126), 139 (128), *157*

B

Baba, S., 91 (129), *101*
Bailey, F. E., 137 (104), *156*
Bailey, G. C., 23 (61), *32*, 126 (69), *155*
Baird, R., 140 (134), *157*
Ballou, E. V., 22 (49), *31*
Barachevsky, V., 25 (66), *32*
Barter, C., 25 (67), *32*, 54 (41), *98*
Barth, R. T., 22 (49), *31*, 120 (55), 127 (55), 134 (55), *155*
Barthomeuf, D., 135 (95), *156*
Basila, M. R., 27 (85), *32*, 60 (52, 53, 55), 61 (52), 62 (52), 63 (52), *99*
Basinski, S., 139 (120), 141 (120, 141, 142, 145), *157*
Bassi, G., 88 (122), *101*
Bazant, V., 106 (13), *153*
Bell, R. P., 1 (1), *3*, 110 (32), 132 (32), *154*, 132 (86), *156*
Belil, G., 140 (135), *157*
Benbenek, S., 141 (144), *157*
Ben-Dor, L., 23 (56), *32*, 88 (123), *101*
Bender, M., 152 (170), *158*
Benesi, H. A., 7 (6), *30*, 14 (31), *31*, 73 (88), 94 (88), *99*, 78 (101), *100*, 113 (38), *154*, 130 (83), *156*
Benson, J. E., 22 (47), *31*
Benson, S. W., 12 (18), 20 (18), 22 (18), *30*, 104 (4), *153*
Beranek, L., 106 (13), *153*
Bergmann, E. D., 139 (125), *157*
Bielanski, A., 36 (7), 37 (7), *43*
Bizreh, Y. W., 130 (82), *155*
Bjerrum, J., 1 (7), *3*
Blackall, E. L., 146 (151), *158*
Blackburn, D. M., 12 (23), 20 (23), 22 (23), *31*, 124 (59), *155*
Boedecker, E. R., 20 (39), 22 (39), *31*, 104 (5), *153*
Bolam, T. R., 96 (142), *101*
Borisova, M. S., 68 (83), *99*
Boudart, M., 22 (47), *31*
Boyle, J. E., 74 (89), *99*
Braun, W., 26 (83), *32*
Breader, D. L., 28 (92), *33*
Bremer, H., 11 (17), *30*, 67 (82), 68 (82), *99*, 132 (85), *156*
Bridges, J. M., 124 (62), *155*
Brønsted, J. N., 1 (3), *3*
Brouwer, D. M., 25 (68), *32*, 124 (61), 127 (71), 147 (61), *155*
Brown, H. C., 135 (99), *156*
Brown, J. F., Jr., 145 (150), *158*
Bulgakov, O. V., 48 (12), *97*, 129 (78), *155*

Bulgakova, Yu. O., 123 (57), 135 (57), *155*

C

Castellá, J., 140 (135), *157*
Castells, J., 140 (135, 137), *157*
Chapman, I. D., 54 (39), *98*, 129 (76), *155*
Chenier, J. H. B., 22 (50), *31*
Cheselske, F. J., 61 (56), *99*
Chessick, J. J., 12 (19), 20 (19), 22 (19), *30*
Christner, L. G., 78 (103), *100*
Clark, A., 12 (23), 20 (23), 22 (23), *31*, 23 (61), *32*, 48 (11), *97*, 124 (59), 126 (69), *155*, 138 (116), *157*
Coing-Boyat, J., 88 (122), *101*
Cook, D., 26 (80), *32*
Corett, R., 139 (125), *157*
Cornelius, E. B., 50 (22), *98*
Custers, J. F. H., 37 (12), *43*, 93 (134), *101*
Cvetanović, R. J., 22 (50), *31*, 138 (117), *157*

D

Danforth, J. D., 23 (54), *31*, 24 (62), *32*, 62 (69), *99*
Dardan, M., 147 (155), *158*
de Boer, J. H., 37 (12, 13), *43*, 93 (134, 135, 136), *101*
De Kimpe, C., 63 (76), *99*
de Mourgues, L., 17 (35), *31*, 23 (60), 25 (73), *32*, 45 (5), *97*, 135 (95), *156*
de Rosset, A. J., 12 (24), 22 (24), *31*
Deyrup, A. J., 5 (2), *30*
Dobrokhotova, N. A., 17 (36), 19 (36), 25 (36), *31*
Dolman, D., 36 (4), *43*
Douglas, G. H., 140 (138), *157*
Dowden, D. A., 72 (87), *99*
Dzisko, V. A., 8 (12), *30*, 66 (81), 67 (81), 68 (83), 70 (85), *99*, 107 (25), 122 (25), *154*, 123 (57), 135 (57), *155*

E

Eastham, A. M., 146 (151), *158*
Eberly, P. E., Jr., 76 (95), 78 (95), *100*
Echigoya, E., 20 (41), 22 (41), *31*, 45 (1, 8), 50 (21), *97*, 61 (58), 68 (84), 70 (84), *99*, 106 (17), *153*, 152 (169), *158*
Eguchi, T., 62 (63), *99*
Eischens, R. R., 27 (84), *32*, 54 (37), *98*
Ellington, P. S., 140 (138), *157*
Emmett, P. H., 62 (73), *99*
Enomoto, S., 134 (92), *156*
Ersini, L., 56 (48), 72 (48), *98*, 107 (26), *154*
Etherington, R. W., 139 (126), *157*
Evans, A. G., 92 (130), *101*

F

Figueras, F., 135 (95), *156*
Filimonov, V. N., 54 (38), *98*
Finch, J. N., 138 (116), *157*
Finstron, C. G., 12 (24), 22 (24), *31*
Fisher, J. B., 105 (8), *153*
Fitzpatrick, J. T., 137 (104), *156*
Fletcher, G. A., 140 (137), *157*
Flockhart, B. D., 150 (164), 151 (165), *158*
Flood, S.H., 136 (100), *156*
Flood, H., 1 (9), *3*
Fokina, E. A., 37 (16), *44*, 51 (23), 55 (23), 52 (28), *98*, 137 (105), *156*, 139 (133), 145 (133), 141 (140), *157*
Förland, T., 1 (9), *3*
Foster, N. F., 138 (117), *157*
Franklin, E. C., 1 (2), *3*
Frei, R., 37 (9), *43*, 53 (31), *98*
Fripiat, J. J., 63 (76), *99*
Frumkin, A., 96 (144), 97 (145), *101*
Fujita, K., 147 (158), *158*
Fukuda, Y., 15 (32), 16 (32), *31*, 134 (90), *156*
Fukui, K., 90 (128), *101*, 115 (43), *154*, 137 (107), *156*, 139 (122), *157*
Furukawa, J., 26 (82), *32*

G

Gaaf, J., 127 (72), *155*
Garves, K., 88 (125), *101*
Gehlen, H., 1 (6), *3*
Gerberich, H. R., 48 (10), 74 (10), *97*, 63 (77), 64 (77), *99*, 135 (94), *156*
Gergel, W. C., 139 (124), *157*
Germann, A. F. O., 1 (4), *3*
Gil-Av, E., 138 (118), *157*
Giron, J., 149 (161), *158*
Good, G. M., 28 (93), *33*, 62 (71), *99*

Goyce, R. N., 137 (108), *156*
Greensfelder, B. S., 28 (93), *33*, 62 (71), *99*
Grenhall, A., 23 (57), *32*
Gruia, M., 147 (155), *158*

H

Haag, W. O., 13 (26), 26 (26), *31*, 45 (7), 47 (7), *97*, 106 (12), 123 (12), 124 (12), 134 (12), *153*, 138 (114), *157*
Hair, M. L., 48 (14), *97*, 54 (39), *98*, 129 (76), *155*
Haldeman, R. G., 62 (73), *99*
Hall, W. K., 25 (65, 74), 26 (78), *32*, 48 (10), 74 (10), *97*, 61 (56), 63 (77), 64 (77), *99*, 78 (100, 103, 104, 105), *100*, 135 (94), *156*
Halpern, J., 146 (152), *158*
Hamamoto, Y., 62 (63), *99*, 94 (138), *101*, 118 (50), 119 (51), *155*
Hammett, L. P., 5 (2), *30*
Hansford, R. C., 62 (64), 63 (64), 66 (64), 68 (64), *99*, 127 (75), *155*
Hantsche, H., 139 (132), *157*
Hara, N., 7 (5), *30*, 104 (3), *153*
Harmsworth, B. J., 53 (35), *98*
Hasegawa, S., 25 (71, 72), *32*, 52 (26), 56 (46), *98*, 64 (79), *99*, 81 (112), 84 (112), *100*, 139 (121), 142 (121), *157*
Hata, S., 64 (79), *99*
Hattori, H., 7 (9), *30*, 25 (76), *32*, 53 (36), 55 (44), 56 (44), 72 (44), *98*, 58 (51), 59 (51), *99*, 75 (93), 78 (99), 87 (121), *100*, 106 (20), 129 (20), *153*, 127 (74), *155*
Hattori, K., 23 (53), *31*, 84 (113), *100*, 113 (36), 125 (36), *154*
Hayakawa, S., 64 (79), *99*
Heunisch, M., 139 (131), *157*
Higuchi, I., 9 (14), *30*
Hill, F. N., 137 (104), *156*
Hindin, S. G., 28 (94), *33*, 48 (17), *97*, 62 (72, 74), *99*
Hirota, K., 106 (18), *153*, 123 (58), *155*
Hirschler, A. E., 7 (3), *30*, 17 (34), *31*, 25 (77), *32*, 45 (6), 62 (6), 63 (6), 65 (6), *97*, 56 (47), 70 (47), 72 (47), *98*, 74 (91), 75 (91),*100*, 127 (73), *155*
Hobson, M. C., Jr., 8 (10), *30*
Hoffmann, A. K., 140 (139), *157*
Holm, V. C. F., 12 (23), 20 (23), 22 (23), *31*, 23 (61), *32*, 48 (11), *97*, 124 (59), 126 (69), *155*
Holmogorov, V., 25 (66), *32*
Honda, H., 54 (40), *98*, 104 (7), *153*, 129 (77), *155*
Horiuti, J., 147 (154), *158*
Houben, G. M. M., 37 (13), *43*, 93 (135), *101*
Hsieh, P. L., 12 (21), 13 (21), 22 (21), *30*, 124 (60), *155*
Hudson, J. O., 25 (77), *32*, 127 (73), *155*
Hughes, T. R., 28 (87), *32*, 78 (98), *100*
Huston, J. L., 1 (8), *3*

I

Ichikawa, I., 7 (9), *30*, 55 (44), 56 (44), 72 (44), *98*
Ichikawa, M., 151 (167), *158*
Igarashi, K., 45 (3), 48 (3), *97*
Iizuka, T., 51 (24), 53 (36), *98*, 85 (118), *100*
Ikebe, K., 7 (5), *30*, 104 (3), *153*
Ikoma, H., 84 (115), 86 (115), *100*
Imai, H., 26 (82), *32*, 105 (9), *153*
Imanaka, T., 123 (58), *155*
Imelik, B., 25 (70), *32*, 37 (15), *44*
Inowe, M., 134 (92), *156*
Ionescu, A., 147 (155), *158*
Isa, I., 135 (96), *156*
Ishibashi, T., 23 (53), *31*, 84 (113), *100*, 113 (36), 125 (36), *154*
Ishikawa, T., 139 (123), *157*
Ishiya, C., 7 (9), *30*, 55 (44), 56 (44), 72 (44), *98*
Ito, M., 45 (3), 48 (3), *97*, 89 (127), *101*, 115 (45), 125 (45), *154*
Izumi, Y., 66 (80), 68 (80), *99*, 107 (23), 136 (23), *154*

J

Jedrzejewski, H., 141 (141), *157*
Johnson, M. F. L., 28 (95), *33*
Johnson, O., 14 (30), 17 (30), *31*, 103 (2), 113 (2), *153*
Johnson, R. E., 1 (8), *3*
Jordan, W., 37 (11), *43*
Jungreis, E., 23 (56), *32*

K

Kado, S., 39 (18), *44*
Kagiya, T., 90 (128), *101*, 115 (43), *154*, 137 (107), *156*

Kamio, T., 139 (123), *157*
Kanetaka, S., 92 (132), *101*
Kantner, T. R., 27 (85), *32*, 60 (52, 53, 55), 61 (52), 62 (52), 63 (52), *99*
Karakchiev, L. G., 8 (12), *30*, 68 (83), 70 (85), *99*
Kaseda, K., 25 (72), *32*, 56 (46), *98*
Kashiki, I., 133 (89), *156*
Katayama, M., 14 (29), *31*, 36 (5), *43*, 55 (45), 56 (45), 91 (45), *98*, 80 (110), 84 (110), *100*
Kawakami, T., 82 (114, 117), 84 (114), 85 (117), *100*, 91 (129), *101*, 115 (42), 125 (42), *154*
Kawaguchi, T., 25 (71, 72), *32*, 52 (26, 27), 56 (46), *98*, 81 (112), 84 (112), *100*, 139 (121), 142 (121), *157*
Kearby, K., 115 (44), *154*
Keii, T., 92 (132), *101*
Kemball, C., 126 (68), 127 (68), *155*
Kevorkian, V., 12 (22), 22 (22), *31*
Khripin, L. A., 68 (83), *99*
Kholmogorov, V. E., 53 (33, 34), *98*
Kiewlicz, W., 141 (143, 145), *157*
Kiji, J., 106 (18), *153*
Kikuchi, N., 36 (2), 39 (2), *43*, 52 (25), *98*, 144 (148), *158*
Kimura, K., 28 (90), *32*, 129 (80), 132 (80), *155*
King, A., 41 (23), *44*, 95 (139), 96 (142), *101*
Kitahara, A., 105 (10), *153*
Klotz, I. M., 153 (171), *158*
Kobayashi, J., 9 (13, 14), *30*
Kobayashi, T., 15 (32), 16 (32), *31*
Kobayashi, Y., 106 (18), *153*, 123 (58), *155*
Koberstein, E., 106 (13), *153*
Kochloffl, K., 106 (13), *153*
Kodratoff, Y., 25 (70), *32*, 37 (15), *44*
Koizuka, J., 84 (115), 86 (115), *100*
Kolovertnova, M., 123 (57), 135 (57), *155*
Kominami, N., 133 (88), *156*
Kondo, N., 37 (11), *43*
Konoshita, T., 39 (20), *44*, 48 (15), 49 (15), *97*, 143 (147), *158*
Kortüm, G., 26 (83), *32*, 37 (8), *43*, 53 (30), *98*, 89 (126), *101*
Kotanigawa, T., 134 (93), *156*
Kotov, E., 25 (66), *32*
Kotsarenko, N. S., 8 (12), *30*, 68 (83), 70 (85), *99*
Kral, H., 41 (24), *44*, 48 (13), *97*, 150 (163), *158*
Kraus, L. M., 106 (13), *153*
Krieger, K. A., 151 (168), *158*
Krylov, O. V., 37 (16), *44*, 51 (23), 52 (28), 55 (23), *98*, 137 (103, 105), *156*, 139 (133), 145 (133), 141 (140), *157*
Kuhn, S. J., 136 (100), *156*
Kubo, K., 15 (32), 16 (32), *31*
Kubokawa, Y., 21 (45), 22 (45), *31*
Kuki, H., 39 (18), *44*
Kurita, M., 81 (112), 84 (112), *100*
Kurita, S., 25 (72), *32*, 56 (46), *98*
Kushnerev, M. J., 137 (105), *156*

L

Lappert, M. F., 26 (81), *32*
Larson, L. G., 78 (104), *100*
Latremouille, G. A., 146 (151), *158*
Lee, J. K., 62 (75), *99*
Leftin, H. P., 8 (10), *30*, 25 (74), *32*, 61 (56), *99*
Leith, I. R., 150 (164), *158*
Leonard, L., 63 (76), *99*
Lewis, G. N., 1 (5), *3*
Liberti, G., 56 (48), 72 (48), *98*, 107 (26), *154*
Lidwell, O. M., 110 (32), 132 (32), *154*
Liengme, B. V., 78 (100), *100*
Lipinski, Z., 141 (141), *157*
Lodin, V. Y., 53 (33), *98*
Longwell, J. P., 28 (92), *33*
Lucchesi, P. J., 28 (92), *33*
Lutinski, F. E., 48 (10), 74 (10), *97*, 63 (77), 64 (77), *99*
Lux, H., 1 (9), *3*

M

MacIver, D. S., 120 (55), 127 (55), 134 (55), 124 (62), *155*
Mackor, E. L., 127 (72), *155*
Mähl, K. A., 23 (59), *32*
Makarov, A. D., 68 (83), *99*
Malinowski, S., 24 (63), *32*, 36 (1, 7), 37 (7), 40 (1), *43*, 42 (25), *44*, 53 (29), *98*, 139 (120), 141 (120, 141, 142, 143, 144, 145), *157*
Manassen, J., 48 (16), *97*, 148 (159), *158*
Mansfield, R. C., 139 (127), *157*
Mapes, J. E., 27 (84), *32*, 54 (37), *98*
Markova, Z. A., 52 (28), *98*, 137 (105), *156*
Margalith, R., 88 (123), *101*

Matsuda, K., 135 (97), *156*
Matsuda, M., 133 (87), *156*
Matsui, T., 86 (120), *100*
Matsumoto, Y., 106 (21), *154*
Matsumura, S., 92 (133), *101*
Matsuura, K., 45 (3), 48 (3), *97*
Matsuzaki, I., 7 (9), *30*, 15 (32), 16 (32), *31*, 55 (44), 56 (44), 72 (44), *98*, 134 (90), 135 (96), *156*
McCarter, M., 37 (9), *43*, 53 (31), *98*
Meakins, G. D., 140 (138), *157*
Medvedovskii, V., 97 (145), *101*
Meininch, E., 106 (15), *153*
Melik, J. S., 28 (95), *33*
Mestres, R., 140 (135), *157*
Mikovsky, R. J., 62 (62), *99*
Milliken, T. H., Jr., 20 (40), 23 (40, 52), *31*, 50 (22), *98*, 62 (68), *99*, 104 (6), *153*
Mills, G. A., 20 (39), 22 (39), 23 (52), *31*, 28 (94), *33*, 50 (22), *98*, 62 (72, 74), *99*, 104 (5), *153*
Mills, G. H., 20 (40), 23 (40), *31*, 62 (68), *99*, 104 (6), *153*
Mimura, M., 21 (46), 22 (46), *31*
Misono, M., 12 (25), 22 (25), *31*, 119 (52), 122 (52), 120 (56), *155*, 147 (156), 148 (156), *158*
Mita, K., 7 (5), *30*, 104 (3), *153*
Mitsutani, A., 62 (63), *99*, 94 (138), *101*, 118 (50), 119 (51), *155*
Miyamura, A., 147 (158), *158*
Miyashita, S., 87 (121), *100*
Mizushima, T., 105 (10, 11), *153*
Mizutori, T., 64 (79), *99*
Mochida, I., 139 (130), *157*
Mori, K., 39 (19), *44*
Morikawa, K., 105 (10, 11), *153*
Morikawa, S., 52 (26, 27), *98*, 139 (121), 142 (121), *157*
Morinari, E., 54 (43), *98*
Morita, Y., 106 (21), *154*
Moskovskaya, I. F., 17 (36), 19 (36), 25 (36), *31*
Moszezenska, J., 141 (141), *157*
Mugiya, C., 80 (111), 81 (111), 84 (111), *100*, 109 (29), 110 (29), 125 (29), *154*
Mukaida, K., 21 (46), 22 (46), *31*
Murakami, Y., 20 (44), 21 (44), 22 (44, 51), *31*, 107 (22), *154*

N

Naccache, C., 25 (70), *32*, 37 (15), *44*
Nagata, A., 147 (157), *158*
Nakamura, A., 25 (75), *32*
Nakamura, Y., 25 (71), *32*
Nakano, M., 92 (133), *101*
Naruko, E., 41 (21, 22), *44*, 95 (140), 96 (141), 97 (140), *101*, 143 (146), *157*
Nath, J., 96 (142), *101*
Nelson, R. L., 53 (35), *98*
Neufang, O., 149 (161), *158*
Nicolescu, A., 147 (155), *158*
Nicolescu, L. V., 147 (155), *158*
Niiyama, H., 61 (58), 68 (84), 70 (84), *99*, 152 (169), *158*
Nishimura, K., 36 (6), *43*, 56 (49), 58 (49), 84 (49), 91 (49), *98*
Nishizawa, T., 75 (93), *100*, 106 (20), 129 (20), *153*
Nitta, M., 135 (96), *156*
Niwa, K., 39 (18), *44*
Noller, H., 139 (131, 132), *157*, 149 (161), 150 (162), *158*
Noro, K., 84 (116), *100*, 112 (35), 125 (35), *154*
Norris, T. H., 1 (8), *3*
Norton, C. J., 74 (92), *100*, 124 (63), *155*
Nozaki, H., 22 (51), *31*, 107 (22), *154*

O

Oblad, A. G., 20 (39, 40), 22 (39), 23 (52), *31*, 28 (94), *33*, 50 (22), *98*, 62 (68, 72), *99*, 104 (5, 6), *153*
Ogasawara, M., 85 (118), *100*
Ogata, N., 137 (109), *156*
Ogino, Y., 30 (96), *33*, 82 (114, 117), 84 (114), 85 (117), *100*, 91 (129), *101*, 108 (27), 115 (41, 42), 125 (27, 41, 42), *154*
Ohki, K., 147 (158), *158*
Ohnishi, R., 7 (7), *30*, 80 (109), 84 (109), 86 (120), *100*, 108 (28), 125 (28), *154*, 125 (66), *155*, 133 (89), *156*
Ohnishi, T., 151 (167), *158*
Ohno, Y., 53 (36), *98*
Ohta, N., 23 (55), *32*
Ohtsuka, H., 118 (48), *155*
Okada, M., 61 (57), 64 (79), *99*, 105 (10, 11), *153*
Okazaki, S., 114 (40), 115 (40), *154*
Okubo, T., 105 (10), *153*
Okuda, N., 24 (64), 28 (64), *32*
Olah, G. A., 136 (100), *156*
Olszewska, M., 141 (142), *157*

O'Reilly, D. E., 61 (56), *99*
Osanami, W., 91 (129), *101*
Otoma, S., 74 (90), 75 (90), *100*
Ozaki, A., 28 (90), *32*, 88 (124), *101*, 116 (46, 47), *154*, 129 (80), 132 (80), *155*, 147 (158), *158*
Ozaki, S., 114 (40), 115 (40), 125 (40), *154*

P

Panchenkov, G. M., 62 (61), *99*
Parravano, G., 54 (43), *98*
Parry, E. P., 27 (86), *32*, 47 (9), 54 (9), 60 (9), *97*
Pascual, J., 140 (135), *157*
Pasynktewicz, I., 141 (144), *157*
Pearson, R. G., 1 (11), *3*
Pernicone, N., 56 (48), 72 (48), *98*, 107 (26), *154*
Peri, J. B., 28 (89), *32*, 49 (18, 19, 20), *97*, 63 (78), 64 (78), *99*, 106 (16), *153*, 129 (79), *155*, 151 (166), *158*
Perozo, S., 150 (162), *158*
Perrin, M., 17 (35), *31*, 23 (60), 25 (73), *32*, 45 (5), *97*, 135 (95), *156*
Pickert, P. E., 74 (89), *99*
Piguzova, L. I., 78 (97), *100*, 130 (82), *155*
Pillai, C. N., 148 (160), *158*
Pimenov, Y. D., 53 (34), *98*
Pink, R. C., 25 (69), *32*, 37 (15), *44*, 127 (70), *155*, 150 (164), 151 (165), *158*
Pines, H., 13 (26), 26 (26), *31*, 28 (91), *32*, 45 (7), 47 (7), 48 (16), *97*, 106 (12), 123 (12), 124 (12), 134 (12), 106 (14), *153*, 138 (113, 114, 115), 139 (113), *157*, 148 (159, 160), *158*
Plank, C. J., 23 (58), *32*, 62 (59, 70), 66 (70), *99*
Plesch, P. H., 92 (130), *101*
Pliskin, W. A., 27 (84), *32*, 54 (37), *98*
Polanyi, M., 92 (130), *101*
Porter, R. P., 26 (78), *32*, 78 (105), *100*
Prettre, M., 135 (95), *156*
Puri, B. R., 96 (142), *101*

R

Rabo, J. A., 74 (89), *99*, 80 (108), *100*
Rase, H. F., 12 (20), 20 (20), 22 (20), *30*
Rav-Acha, Ch., 138 (118), *157*
Ravoire, J., 106 (14), *153*
Rhee, K. H., 60 (52), 61 (52), 62 (52), 63 (52), *99*
Richardson, J. T., 76 (94), 100
Richardson, R. L., 12 (18), 20 (18), 22 (18), *30*, 104 (4), *153*
Roberts, R. M., 25 (67), *32*
Roev, L. M., 54 (38), *98*
Romanovsky, B. V., 78 (97), *100*, 130 (82), *155*
Rooney, J. J., 25 (69), *32*, 126 (68), 127 (68, 70), *155*
Roper, E. E., 11 (16), *30*
Rosolovskaja, E. N., 106 (19), 123 (19), *153*

S

Saegusa, T., 26 (82), *32*
Sakurai, C., 133 (88), *156*
Saito, K., 38 (17), *44*, 144 (149), *158*
Saito, Y., 12 (25), 22 (25), *31*, 26 (79), *32*, 120 (56), *155*, 139 (130), *157*
Sano, K., 25 (75), *32*
Sano, M., 90 (128), *101*
Sano, T., 115 (43), *154*, 135 (98), 137 (107), *156*
Sansoni, B., 1 (7), *3*
Sasaki, H., 92 (133), *101*
Sato, M., 13 (27), *31*, 25 (76), *32*, 58 (51), 59 (51), *99*, 124 (64), 126 (64), 127 (64, 74), *155*, 131 (84), *156*
Sauer, R. W., 151 (168), *158*
Schächter, Y., 138 (115), 138 (118), *157*
Schaffer, A. D., 37 (10), *43*, 53 (32), *98*
Schaffer, P. C., 78 (102), *100*
Schatunowskaja, H., 96 (142), *101*
Scheidt, F. M., 139 (119), *157*
Schmidle, C. J., 139 (127), *157*
Schmitz, E., 139 (131), *157*, 149 (161), 150 (162), *158*
Schomaker, V., 80 (108), *100*
Schwab, G.-M., 41 (24), *44*, 48 (13), *97*, 88 (125), *101*, 150 (163), *158*
Scott, J. A. N., 151 (165), *158*
Sebba, F., 105 (8), *153*
Serain, S., 149 (161), *158*
Serratosa, F., 140 (135, 136), *157*
Shaap, L. A., 138 (113), 139 (113), *157*
Shakhnovskaja, O. L., 106 (19), 123 (19), *153*
Sharma, L. R., 96 (142), *101*
Shatenshtein, A. I., 1 (10), *3*

Sheidt, F. M., 139 (119), *157*
Shephard, F. E., 126 (68), 127 (68), *155*
Shiba, T., 13 (27), 20 (44), 21 (44), 22 (44), *31*, 25 (75, 76), *32*, 58 (51), 59 (51), 66 (80), 68 (80), *99*, 75 (93), 78 (99), *100*, 106 (20), 129 (20), *153*, 107 (23), 136 (23), *154*, 124 (64), 126 (64), 127 (64, 74), *155*, 131 (84), *156*
Shilov, N., 96 (142, 143), *101*
Shimizu, H., 45 (2), *97*
Shimidzu, T., 90 (128), *101*, 115 (43), *154*
Shimomura, K., 39 (20), *44*, 48 (15), 49 (15), *97*, 143 (147), *158*
Shirasaki, T., 21 (46), 22 (46), *31*, 61 (57), 64 (79), *99*, 105 (10, 11), *153*
Shlygin, A., 97 (145), *101*
Sibbert, D. J., 62 (59), *99*
Singh, D. D., 96 (142), *101*
Skinner, B. G., 132 (86), *156*
Skinner, H. A., 92 (130), *101*
Sloczynski, J., 36 (7), 37 (7), *43*, 42 (25), *44*, 53 (29), *98*
Smith, A., 17 (34), *31*, 56 (47), 70 (47), 72 (47), *98*
Smith, R. B., 62 (59), *99*
Smith, R. K., 146 (153), *158*
Soltys, E., 141 (143), *157*
Soma, J., 53 (36), *98*
Soma, M., 151 (167), *158*
Stamires, D. N., 74 (89), *99*, 78 (106), 79 (106), *100*
Steinberg, K. H., 11 (17), *30*, 67 (82), 68 (82), *99*, 132 (85), *156*
Steiner, R. O., 12 (22), 22 (22), *31*
Stewart, R., 36 (4), *43*
Stock, L. M., 135 (99), *156*
Stone, F. S., 22 (48), *31*
Stone, H., 25 (67), *32*
Stone, R. L., 12 (20), 20 (20), 22 (20), *30*
Stright, P., 23 (54), *31*
Suganuma, F., 61 (57), *99*
Suzuki, A., 45 (3), 48 (3), *97*, 89 (127), *101*, 115 (45), 125 (45), *154*
Suzuki, H., 52 (26), *98*, 139 (121), 142 (121), *157*
Suzuki, S., 63 (76), *99*
Swain, C. G., 145 (150), *158*
Swindells, R., 140 (138), *157*
Szczepanska, S., 24 (63), *32*, 36 (1, 7), 37 (7), 40 (1), *43*, 42 (25), *44*, 53 (29), *98*, 141 (145), *157*

T

Tachibana, T., 24 (64), 28 (64), *32*
Tada, A., 89 (127), *101*, 115 (45), 125 (45), *154*
Takagi, T., 92 (132), *101*
Take, J., 36 (2), 39 (2), *43*, 52 (25), *98*, 139 (130), *157*, 144 (148), *158*
Takei, M., 139 (122), *157*
Takemura, K., 25 (75), *32*
Takeshita, T., 86 (120), *100*, 109 (30), 125 (30), *154*, 125 (66, 67), *155*, 133 (89), 135 (98), *156*, 147 (157), *158*
Takeuchi, T., 20 (43), 22 (43), *31*
Takida, H., 84 (116), *100*, 112 (35), *154*
Tamaru, K., 88 (124), *101*, 151 (167), *158*
Tamele, M. W., 13 (28), *31*, 62 (67), 63 (67), *99*, 103 (1), 104 (1), *153*
Tanabe, K., 7 (7, 8, 9), *30*, 14 (29), 15 (32), 16 (32), 17 (33, 37), *31*, 36 (3, 5), 38 (3), 43 (3), *43*, 38 (17), 42 (26), *44*, 45 (2), *97*, 51 (24), 53 (36), 54 (42), 55 (44, 45), 56 (44, 45), 58 (42, 50), 62 (50), 72 (44), 91 (45), *98*, 80 (109, 110, 111), 81 (111), 84 (109, 110, 111, 115), 85 (118), 86 (115, 120), 87 (121), *100*, 92 (131), *101*, 108 (28), 109 (29, 30, 31), 110 (29, 33), 112 (34), 113 (37), 114 (37), 119 (37), 125 (28, 29, 30, 37), *154*, 125 (65, 66, 67), 132 (65), 135 (65), *155*, 133 (87, 89), 134 (90), 135 (96, 97, 98), *156*, 144 (149), 147 (157), *158*
Tanaka, K., 88 (124), *101*, 116 (46, 47), *154*
Tani, M., 108 (27), 125 (27), *154*
Taniguchi, S., 54 (40), *98*, 129 (77), *155*
Tarama, K., 23 (53), *31*, 54 (40), *98*, 84 (113), *100*, 104 (7), *153*, 113 (36), 125 (36), *154*, 129 (77), *155*
Temma, M., 20 (42), 22 (42), *31*
Tench, A. J., 53 (35), *98*
Teranishi, S., 23 (53), *31*, 54 (40), *98*, 84 (113), *100*, 104 (7), *153*, 113 (36), 125 (36), *154*, 129 (77), *155*
Terenin, A. N., 8 (11), *30*, 25 (66), *32*, 37 (14), *43*, 53 (33, 34), 54 (38), *98*, 93 (137), *101*
Terlecki-Baricevic, A., 147 (155), *158*
Tezuka, Y., 20 (43), 22 (43), *31*
Thoang, Ho si, 130 (82), *155*
Thomas, C. L., 19 (38), 28 (38), *31*, 62 (66), 63 (66), 68 (66), *99*
Tkhoang, K. S., 78 (97), *100*

Tomlinson, J. W., 1 (9), *3*
Topchieva, K. V., 17 (36), 19 (36), 25 (36), *31*, 78 (97), *100*, 106 (19), 123 (19), *153*, 130 (82), *155*
Tovar, P., 150 (162), *158*
Toyama, S., 21 (45), 22 (45), *31*
Toyoshima, I., 84 (115), 86 (115), *100*
Trambouze, Y., 17 (35), *31*, 23 (60), 25 (73), *32*, 45 (5), *97*, 60 (54), *99*, 135 (95), *156*
Tretiakov, I. I., 52 (28), *98*
Tschmutov, K., 96 (142), *101*
Tsutsumi, S., 62 (65), *99*
Tung, S. E., 106 (15), *153*
Turkevich, J., 22 (51), *31*, 78 (106, 107), 79 (106, 107), *100*, 107 (22), *154*, 146 (153), *158*

U

Uchida, H., 20 (42), 22 (42), *31*, 39 (20), *44*, 48 (15), 49 (15), *97*, 105 (9), *153*, 143 (147), *158*
Uejima, T., 26 (82), *32*
Uematsu, T., 75 (93), *100*, 106 (20), 129 (20), *153*
Ukihashi, H., 74 (90), 75 (90), *100*
Uppal, S. S., 150 (164), *158*
Ushiba, K., 22 (47), *31*
Ussanowitch, M., 1 (6), *3*
Uvarov, A. V., 48 (12), *97*, 129 (78), *155*
Uytterhoven, J. B., 78 (103), *100*

V

Vaughan-Jackson, M. M., 110 (32), 132 (32), *154*
Vinnikova, T. S., 123 (57), 135 (57), *155*
Voge, H. H., 28 (93), *33*, 62 (71), *99*
Vogel, J., 26 (83), *32*, 89 (126), *101*
Voltz, S. E., 17 (34), *31*, 56 (47), 70 (47), 72 (47), *98*, 70 (86), *99*
von E. Doering, W., 140 (139), *157*

W

Wackher, R. C., 28 (91), *32*
Wagner, C. D., 54 (41), *98*
Walling, C., 5 (1), 7 (1), *30*
Ward, J. W., 76 (96), 78 (96), 79 (96), *100*, 127 (75), 130 (81), *155*
Watanabe, Y., 17 (33), *31*, 92 (131), *101*, 107 (24), 110 (33), 113 (37), 114 (37), 119 (37), 125 (24, 37), *154*
Webb, A. N., 11 (15), 19 (15), 22 (15), *30*, 28 (88), *32*, 45 (4), *97*
(15), *31*, 28 (88), *32*, 45 (4), *97*
Weil-Malherbe, H., 7 (4), *30*
Weiss, J., 7 (4), *30*
Weisz, P. B., 62 (62), *99*
Weller, S. W., 48 (17), *97*, 62 (75), 70 (86), *99*, 96 (142), *101*
Whalley, L., 22 (48), *31*
White, H. M., 28 (87), *32*, 78 (98), *100*
Wilmot, W. H., 120 (55), 127 (55), 134 (55), 124 (62), *155*
Wilson, J. H., 96 (142), *101*
Winstein, S., 140 (134), *157*
Wojciechowska, E., 141 (144), *157*

Y

Yamadaya, M., 39 (20), *44*, 48 (15), 49 (15), *97*, 143 (147), *158*
Yamaguchi, S., 62 (65), *99*
Yamaguchi, T., 7 (8), *30*, 17 (37), *31*, 36 (3), 38 (3), 43 (3), *43*, 42 (26), *44*, 45 (2), *97*, 54 (42), 58 (42, 50), 62 (50), *98*, 125 (65), 132 (65), 135 (65), *155*, 135 (97), *156*
Yamamoto, F., 81 (112), 84 (112), *100*
Yamamoto, Y., 89 (127), *101*, 115 (45), 125 (45), *154*
Yamanaka, T., 118 (49), *155*
Yoneda, Y., 12 (25), 22 (25), *31*, 26 (79), *32*, 36 (2), 39 (2), *43*, 52 (25), *98*, 119 (52, 53, 54), 120 (56), 122 (52), *155*, 139 (130), *157*, 144 (148), 147 (156), 148 (156), *158*
Yoshida, K., 25 (76), *32*, 58 (51), 59 (51), *99*, 127 (74), *155*
Yoshida, S., 54 (40), *98*, 104 (7), *153*, 129 (77), *155*
Yoshii, Y., 45 (3), 48 (3), *97*
Yoshioka, T., 84 (115), 86 (115), *100*
Yound, T. F., 96 (142), *101*
Yumoto, S., 137 (109), *156*

Z

Zaugg, H. E., 37 (10), *43*, 53 (32), *98*
Zeitlin, H., 37 (9, 11), *43*, 53 (31), *98*
Zettlemoyer, A. C., 12 (19), 20 (19), 22 (19), *30*
Zieleniewska, H., 141 (142), *157*
Zusman, R. J., 68 (83), *99*

Subject Index

Catalysts are listed separately in the Catalyst Index (p.174)

A

Acetonitrile, 141
Acid-base bifunctional catalysis, 146, 152
Acidic properties, 5
  Acid amount, 13
  Acid strength, 5
  Acidity, 13
Acrolein, 139
Acrylonitrile, 139
Addition reaction
  of epoxide, 139
  of nitromethane, 139
Adsorption, heat of, 12
Adsorption method, 10
Aldol condensation, 142
Alkylation, 136
Allyl anion, 137
Amine titration method, 13
Aminoazobenzene, 8
2-Amino-5-azotoluene, 6
Ammonia, 20, 27
Anion exchange resin, 139
Anthracene, 25, 127
Anthraquinone, 6

B

Basic properties, 35
  Base amount, 38
  Basicity, 38
  Basic strength, 35, 145
Bathochromic shift, 93
Bayerite, 123
Beckmann rearrangement of cyclohexanone oxime, 107, 125
Benzalacetophenone, 6, 8
Benzaldehyde, 144
Benzeneazodiphenylamine, 6
p-Benzoyldiphenyl, 8
Benzyl benzoate, 144
Benzyl chloride, 135
Boehmite, 123
Brønsted acid, 23, 63, 87, 127
  Brønsted acidity, 59, 131
  Brønsted acid site, 48, 63, 65, 77, 125
Bromobenzene, 135
2-Bromo-2-butene, 150
1-Bromo-2-methyl-1-propene, 150
Bromocresol purple, 36
Bromophenol blue, 36
cis-Bromostilbene, 150
Bromothymol blue, 36
n-Butylamine, 13, 18, 20
Butter yellow, 6

C

Calorimetric titration, 19, 42
Cannizzaro reaction, 139
Carbenes, 140
Carbanion, 137
Cation radical, 24
Charcoal, 42, 95
4-Chloroaniline, 36
4-Chloro-2-nitroaniline, 36
3-Chloro-2,4,6-trinitroaniline, 8
Concerted mechanism, 146
Condensation, 139
  of acetaldehyde, 139, 142
  of glucose, 125
  of formaldehyde, 139
  of propionaldehyde, 139
Conversion of sucrose, 23
Cracking activity of cumene, 127, 129
Cracking
  of cumene, 13, 62, 106, 107, 124, 150
  of diisobutylene, 122
  of *n*-octane, 124
Crystal violet, 6, 26
Cyclohexane, 47, 140

D

Dealkylation of *tert*-butylbenzene, 122
Decomposition
  of *n*-butyl alcohol, 123
  of cetane, 23
  of cumene, 103
  of diisobutylene, 106
  of formic acid, 116
  of isobutane, 75, 106, 127
  of hydrogen peroxide, 41
Dehydration
  of alcohols, 107
  of alkylcyclohexanols, 148
  of bornanols, 148
  of *tert*-butanol, 120
  of *n*-butyl alcohol, 135
  of cyclohexanol, 85
  of decalols, 148
  of isopropanol, 115, 125
  of isopropyl alcohol, 107, 115, 122, 141
  of menthols, 148
  of 4-methyl-2-pentanol, 135
  of neomenthols, 148
  of 2-phenyl-1-propanol, 148
Dehydrochlorination of 1,1,2-trichloroethane, 139, 143
Dehydrogenation of isopropyl alcohol, 141
Dehydrohalogenation, 139
Depolymerization of paraldehyde, 109, 125, 132
2,3-Dibromobutane, 149
2,3-Dichlorobutane, 139, 149
1,2-Dichloro-2,2-diphenylethane, 139
2,4-Dichloro-6-nitroaniline, 8
Dicinnamalacetone, 6
Diethyl carbonate, 151
Differential heat of adsorption, 12, 124
Differential surface entropy, 123
Differential thermal analysis (DTA) method, 11, 21
7,7-Dihalobicyclo [4.1.0] heptane, 140
Dimerization
  of propylene, 122
  of isobutane, 122
p-Dimethylaminoazobenzene 6, 14
4-Dimethylaminoazo-1-naphthalene, 6
4,4-Dimethyl-1,3-dioxane, 116
Dimethyl yellow, 6, 14
2,4-Dinitroaniline, 8, 36
2,4-Dinitrotoluene, 8
Dioxane, 17, 25
1,1-Diphenyl-2-bromoethylene, 150
Diphenylcarbinol, 8
Diphenylethylene, 28
Diphenylpicrylhydrazyl (DPPH), 25
Disproportionation,
  of fluorochloromethane, 125
  of toluene, 105, 130, 136
  of trichlorofluoromethane, 114

E

Electronegativity, 88, 115
Esterification of phthallic acid, 23, 125
Ethyl acetate, 17, 25, 26
Ethylenediamine, 71
Exchange method, 41

F

Furfural, 139
β 2-Furylacrolein, formation of, 139

H

Hammett acidity function, 5
Hammett indicators, 8
H_R indicator, 6, 75, 129
Hydration of propylene, 115, 125
Hydrogen switch mechanism, 146
Hydrogen transfer, 124
Hydrolysis,
  of cholestan-3β-yl 3,5-dinitrobenzoate, 140
  of diethyl carbonate, 151
  of methylene chloride, 133
  of toluenesulfonate, 140
Hydrotactoids, 153
2-Hydroxypridine, 145

I

Isobutylene, 118
Isomerization, 136
  of allybenzene, 123
  of butene, 62
  of 1-butene, 118, 123, 129, 146
  of *n*-butene, 122
  of crotononitriles, 147
  of cyclohexane, 13
  of dimethyl-1-butene, 13
  of 3,3-dimethyl-1-butene, 127
  of isobutylene, 122
  of limonene, 138
  of *o*-oxylene, 119, 124
  of 1-pentene, 127, 138
  of α-pinene, 125, 132

of *o*-xylene, 127

K

Knoevenagel reaction, 139

L

Lewis acid, 23, 24, 63, 87, 127
Lewis acidity, 47, 59
Lewis acid site, 48, 63, 77, 125

M

Malachite green, 26
4-Methyl-2-pentene, 138
Methyl red, 6
Michael condensation, 139
Mononitrophenol, 37
Montmorillonite, 7
Mounted acids, 93

N

Neutral red, 6
o-Nitroaniline, 8
p-Nitroaniline, 8, 36
p-Nitroazobenzene, 8
Nitrobenzene, 8
p-Nitrobenzeneazo (*p'*-nitro) diphenylamine, 6
p-Nitrodiphenylamine, 8
o-Nitrophenol, 37
p-Nitrophenol, 37, 93
p-Nitrotoluene, 8
Nitromethane, 139

O

Oligomerization, 119
Oxidation of methanol, 107

P

Perylene, 25, 127
Phenol, 37
Phenolphthalein, 26
Phenylazonaphthylamine, 6, 9
p-Phenylenediamine, 24
Piperidine, 12, 20, 27
Polymerization activity, 136
for ethylene, 106
for propylene, 106
Polymerization
of acetaldehyde, 116
of aldehydes, 112, 125
of ethylene, 105
of ethylene oxide, 115, 137
of formaldehyde, 137
of isobutylene, 106, 116
of isobutylvinylether, 115, 125
of lactam, 137
of β-propiolactone, 137
of propylene, 62, 75, 103, 113, 118, 124, 125, 126
of propylene oxide, 137
Potentiometric titration, 24
Push-pull mechanism, 146
Pyridine, 11, 20, 27
Pyridinium ion, 48, 60
Pyrrole, 20

Q

Quinoline, 20

S

Selectivity, 133
Solid base catalysis, 136
Spectrophotometric method, 7
Spiro [2.5] octa-1,4-dien-3-one, 140
Synchronous mechanism, 146
Synthesis
of isoprene, 116
of methanol, 107, 125

T

Ternary mechanism, 146
Tetracyanoethylene, 37
α–D–Tetramethylglucose, 145
Thymolphthalein, 37
Transalkylation, 107
Trimethylamine, 12, 20
2,4,6-Trimethylbenzyl alcohol, 8
2,4,6-Trinitroaniline 8, 36
Trinitrobenzene, 79
1,3,5-Trinitrobenzene, 37, 53
Triphenylcarbinol, 8
Triphenylcarbonium ions, 25, 127
Triphenylmethane, 25, 127
Toluene, 135

X

Xanthone, 26
m-, *o*-, *p*-Xylene, 135
2,6–Xylenol, 134

Catalyst Index

A

$AlCl_3$, 91, 136
AlF_3, 114
Al_2O_3, 22, 39, 40, 45, 106, 124, 129, 135, 148
  α–Al_2O_3, 136
  γ–Al_2O_3, 123 (fluorided, 127)
  θ–Al_2O_3, 123
  η–Al_2O_3, 47, 67, 136
  χ–Al_2O_3, 136
$Al_2O_3 \cdot Cr_2O_3$, 70, 71
$Al_2O_3 \cdot B_2O_3$, 66, 107, 124, 131
$Al_2O_3 \cdot ThO_2 \cdot H_2SO_4$, 147
$AlPO_4$, 89, 116
$Al_2(SO_4)_3$, 83
$Al_3(SO_4)_3/SiO_2$, 29
As_2O_3, 56

B

BF_3, 41, 48, 127
BPO_4, 89, 116
$BaCO_3$, 36, 91
BaF_2, 93
BaO, 52
BaS, 58, 84
$BaSO_4$, 84
BeO, 37, 51

C

C (active carbon, activated charcoal), 41, 95
C_5H_5NO (2-hydroxypyridine), 145
$CaCO_3$ 91, 138
$CaCl_2$, 91, 114
CaF_2, 91, 93, 114
CaO, 38, 51, 137, 144, 150
$Ca(OH)_2$, 137, 139
$Ca_3(PO_4)_2$, 150
CaS, 58
$CaSO_4$, 84, 150
CeO_2, 56
$Ce_2O \cdot MnO \cdot MgO$, 134
$Ce_2(SO_4)_3 \cdot 8H_2O$, 84
$CdSO_4$, 83, 88
Clay minerals, 73
$CoSO_4$, 84, 88
$CrCl_3$, 91, 114
Cr_2O_3, 56, 57
$Cr_2O_3 \cdot Al_2O_3$, 70, 71
$CrPO_4$, 116
$Cr_2(SO_4)_3$, 84
Cu_2Cl_2, 91
$Cu_3(PO_4)_2$, 89, 116
$CuSO_4$, 80, 84, 109

F

Filtrol, 28
Fluorided γ-alumina, 127
$FeCl_3$, 91, 114
$Fe_2O_3 \cdot ZnO$, 134
$FePO_4$, 89, 116
$FeSO_4$, 84, 88
$Fe(SO_4)_3$, 84

H

HF–Al_2O_3, 11
$HgCl_2$, 91, 114
2-Hydroxypyridine (C_5H_5NO), 145

K

Kaolinite, 7
K/graphite, 138
KBO_2, 149

$KC_{11}H_7O_3$ (Potassium 2-naphthol-3-carboxylate), 147
KCN, 36
K_2CO_3, 36, 91, 139, 149
KH dispersed in mineral oil, 137
$KHCO_3$, 36, 91
KNH_2/Al_2O_3, 137
$KNaCO_3$, 36
KOH/Al_2O_3, 140

L

Li_2CO_3/SiO_2, 139
Li_3PO_4, 139
Li_2SO_4, 150

M

Metal sulfates, 124, 146
$MgCO_3$, 91
$4MgCO_3 \cdot Mg(OH)_2 \cdot 4H_2O$, 134
MgO, 50, 137
$Mg_3(PO_4)_2$, 116
$MgSO_4$, 83, 88
$MnSO_4$, 88
$MnSO_4/SiO_2$, 29
Montmorillonite, 7
MoO_3, 56
$MoO_3 \cdot Fe_2(MoO_4)_3$, 72, 107
Mordenite, 130

N

$(NH_4)_2CO_3$, 36
Na/Al_2O_3, 138
Na/K_2CO_3, 137
$NaC_7H_5O_3$ (sodium salicylate), 147
Na_2CO_3, 36, 91
NaH, 138
$NaOH/SiO_2$ (Na/SiO_2), 24, 40, 139
$Na_2WO_4 \cdot 2H_2O$, 36
$NiO \cdot SiO_2$, 129
$Ni_3(PO_4)_2$, 89, 116
$NiSO_4$, 81, 109, 119

P

$PbCl_2$, 114
PbO, 139
Potassium 2-naphthol-3-carboxylate ($KC_{11}H_7O_3$), 147

S

$SbCl_3$, 91
Silica gel ($SiO_2 \cdot nH_2O$), 29, 54
$SiO_2 \cdot nH_2O$ (Silica gel), 54
$SiO_2 \cdot Al_2O_3$, 19, 20, 22, 28, 43, 58, 61, 62, 103, 104, 105, 119, 124, 127, 129
$SiO_2 \cdot BF_3$, 127
$SiO_2 \cdot BaO$, 68, 69
$SiO_2 \cdot BeO$, 70
$SiO_2 \cdot CaO$, 68, 69
$SiO_2 \cdot CaO \cdot MgO$, 39
$SiO_2 \cdot Ga_2O_3$, 70
$SiO_2 \cdot La_2O_3$, 70
$SiO_2 \cdot MgO$, 11, 22, 28, 66, 69, 107, 120, 127, 131
$SiO_2 \cdot PbO$, 70
$SiO_2 \cdot SnO$, 70
$SiO_2 \cdot SrO$, 68, 69
$SiO_2 \cdot Y_2O_3$, 70
$SiO_2 \cdot ZrO_2$, 22, 66, 107
$SnCl_2$, 91
Sodium salicylate ($NaC_7H_5O_3$), 147
$SrCO_3$, 36, 91
SrO, 51
$SrSO_4$, 84

T

$TiCl_3$, 17, 92
TiO_2, 56
$TiO_2 \cdot ZnO$, 72, 73
$Ti_3(PO_4)_4$, 116

V

V_2O_5, 56

Z

Zeolites, 73, 124, 130
  Zeolite-X, 22, 74, 106
  Zeolite-Y, 22, 74, 76
Ziegler-Natta catalyst, 92
$ZnCl_2$, 91
$Zn_3(PO_4)_2$, 116
ZnO, 55
$ZnO \cdot Cr_2O_3 \cdot SiO_2 \cdot Al_2O_3$, 107
ZnS, 57
$ZnSO_4$, 83, 88
$ZnSO_4/SiO_2$, 29
$Zr_3(PO_4)_4$, 90, 116